Quantum Logic in Algebraic Approach

Fundamental Theories of Physics

An International Book Series on The Fundamental Theories of Physics:
Their Clarification, Development and Application

Editor:
Alwyn van der Merwe, *University of Denver, U.S.A.*

Editoral Advisory Board:
Lawrence P. Horwitz, *Tel-Aviv University, Israel*
Brian D. Josephson, *University of Cambridge, U.K.*
Clive Kilmister, *University of London, U.K.*
Pekka J. Lahti, *University of Turku, Finland*
Günter Ludwig, *Philipps-Universität, Marburg, Germany*
Asher Peres, *Israel Institute of Technology, Israel*
Nathan Rosen, *Israel Institute of Technology, Israel*
Eduard Prugovecki, *University of Toronto, Canada*
Mendel Sachs, *State University of New York at Buffalo, U.S.A.*
Abdus Salam, *International Centre for Theoretical Physics, Trieste, Italy*
Hans-Jürgen Treder, *Zentralinstitut für Astrophysik der Akademie der Wissenschaften,*
　Germany

Quantum Logic in Algebraic Approach

by

Miklós Rédei
Faculty of Natural Sciences,
Loránd Eötvös University,
Budapest, Hungary

KLUWER ACADEMIC PUBLISHERS
DORDRECHT / BOSTON / LONDON

A C.I.P. Catalogue record for this book is available from the Library of Congress.

ISBN 0-7923-4903-2

Published by Kluwer Academic Publishers,
P.O. Box 17, 3300 AA Dordrecht, The Netherlands.

Sold and distributed in the U.S.A. and Canada
by Kluwer Academic Publishers,
101 Philip Drive, Norwell, MA 02061, U.S.A.

In all other countries, sold and distributed
by Kluwer Academic Publishers,
P.O. Box 322, 3300 AH Dordrecht, The Netherlands.

Manuscript translated from the original Russian by V. Kisin.

Printed on acid-free paper

All Rights Reserved
©1998 Kluwer Academic Publishers
No part of the material protected by this copyright notice may be reproduced or
utilized in any form or by any means, electronic or mechanical,
including photocopying, recording or by any information storage and
retrieval system, without written permission from the copyright owner

Printed in the Netherlands.

Table of Contents

Preface   **ix**

1 Introduction   **1**
    1.1 Bibliographic notes   8

2 Observables and states in the Hilbert space formalism of quantum mechanics   **11**
    2.1 Observables   11
    2.2 States   20
    2.3 Bibliographic Notes   27

3 Lattice theoretic notions   **29**
    3.1 Basic notions in lattice theory   29
    3.2 Bibliographic notes   43

4 Hilbert lattice   **45**
    4.1 Hilbert space and the lattice of subspaces   45
    4.2 Subspaces and projections   54
    4.3 Bibliographic notes   60

5 Physical theory in semantic approach   **61**
    5.1 Physical theory as semi-interpreted language   61
    5.2 The logic of classical mechanics   64
    5.3 Hilbert lattice as logic   68
    5.4 Bibliographic notes   74

6 Von Neumann lattices   **77**
    6.1 Von Neumann algebras   77
    6.2 Von Neumann lattices   82
    6.3 Appendix: proofs of propositions related to the classification theory of von Neumann algebras   90
    6.4 Bibliographic notes   100

TABLE OF CONTENTS

7 The Birkhoff-von Neumann concept of quantum logic   **103**
  7.1 Quantum logic as event structure of non-commutative probability 103
      7.1.1 Digression: von Neumann's concept of probability in quantum mechanics in the years 1926-1932 105
      7.1.2 Back to the type II_1 factor 112
  7.2 Probability is logical 113
  7.3 Bibliographic notes 116

8 Quantum conditional and quantum conditional probability   **119**
  8.1 Minimal implicative criteria and quantum conditional . . . 119
  8.2 Conditional probability and statistical inference 127
  8.3 Breakdown of Stalnaker's Thesis in quantum logic 134
  8.4 Bibliographic notes 137

9 The problem of hidden variables   **139**
  9.1 Historical remarks 139
  9.2 Notion of and no-go results on dispersive hidden theories . . 144
      9.2.1 Definition of dispersive hidden theory 144
      9.2.2 Negative results on dispersive hidden theories . . . 149
  9.3 No-go results on entropic hidden theories 156
  9.4 The problem of local hidden variables 160
      9.4.1 Bell's question and Bell's inequality 160
      9.4.2 No-go proposition on dispersive local hidden theories 164
  9.5 Bibliographic notes 169

10 Violation of Bell's inequality in quantum field theory   **171**
  10.1 Basic notions of algebraic quantum field theory 171
  10.2 Bell correlation and Bell's inequality 180
  10.3 Violation of Bell's inequality in quantum field theory . . . 184
  10.4 Superluminal correlations in quantum field theory 188
  10.5 Bibliographic notes 190

11 Independence in quantum logic approach   **191**
  11.1 Logical independence in quantum logic 193
      11.1.1 Logical notions of independence 193
      11.1.2 Logical and statistical independence 197
  11.2 Counterfactual probabilistic independence 204
      11.2.1 Concept of counterfactual probabilistic independence 205
      11.2.2 Counterfactual probabilistic independence in quantum field theory 207
  11.3 Bibliographic notes 213

12 Reichenbach's common cause principle and quantum field theory — **215**
12.1 Reichenbach's common cause principle 216
12.2 Do superluminal correlations have a probabilistic common cause? 219
12.3 Bibliographic notes 224

References — **227**

Index — **235**

Preface

This work has grown out of the lecture notes that were prepared for a series of seminars on some selected topics in quantum logic. The seminars were delivered during the first semester of the 1993/1994 academic year in the Unit for Foundations of Science of the Department of History and Foundations of Mathematics and Science, Faculty of Physics, Utrecht University, The Netherlands, while I was staying in that Unit on a European Community Research Grant, and in the Center for Philosophy of Science, University of Pittsburgh, U.S.A., where I was staying during the 1994/1995 academic year as a Visiting Fellow on a Fulbright Research Grant, and where I also was supported by the István Széchenyi Scholarship Foundation. The financial support provided by these foundations, by the Center for Philosophy of Science and by the European Community is greatly acknowledged, and I wish to thank D. Dieks, the professor of the Foundations Group in Utrecht and G. Massey, the director of the Center for Philosophy of Science in Pittsburgh for making my stay at the respective institutions possible.

I also wish to thank both the members of the Foundations Group in Utrecht, especially D. Dieks, C. Lutz, F. Muller, J. Uffink and P. Vermaas and the participants in the seminars at the Center for Philosophy of Science in Pittsburgh, especially N. Belnap, J. Earman, A. Janis, J. Norton, and J. Forge not only for their interest in the seminars and in the subsequent stimulating discussions but also for their hospitality in Utrecht and Pittsburgh, which made my stay in Utrecht and Pittsburgh a most pleasant experience. Special thanks go to my friend and colleague D. Petz, professor of mathematics in the Mathematics Institute of the Technical University in Budapest, Hungary, who encouraged me to complete the lecture notes and who corrected a number of errors in the manuscript. Needless to say, neither he, nor any of those mentioned here bear any responsibility whatsoever for any errors that might remain in the work.

Part of the writing was done in the summer of 1996 during a stay at the University of Florida, Gainesville, Florida, U.S.A., where I was supported by a grant from the Soros Foundation.

Some of the results presented in this work were obtained in research supported also by the Hungarian National Science Fund (OTKA) under the following contract numbers: 1900, T 015606, T 013853, T 023447 and T 1850. I thank OTKA for these research grants.

Miklós Rédei

Budapest, February 1997

CHAPTER 1

Introduction

The main idea of quantum logic is very simple: Let Q be a selfadjoint operator (defined on the Hilbert space $\mathcal{H}$) representing an observable physical quantity, and let P^Q be its spectral measure. According to quantum mechanics the probability that Q takes its value in the interval d is equal to one if the system is prepared in a state vector ψ that lies in the spectral subspace $P^Q(d)$, a closed, linear subspace in $\mathcal{H}$. One can express this fact by saying that ψ *makes true the proposition* "Q has its value in d with probability one" $(=Prop(Q,d))$. Identifying, as it is common in logic, a proposition with the set of interpretations that make the proposition true, the subspace $P^Q(d)$ can be viewed as the representative of the proposition $Prop(Q,d)$. Given two such propositions $Prop(Q,d)$ and $Prop(Q',d')$ the proposition

$Prop(Q,d) \wedge Prop(Q',d') =$
"Q has value in d and Q' has value in d' with probability one"

is represented by the intersection

$$P^Q(d) \cap P^{Q'}(d')$$

of the subspaces representing $Prop(Q,d)$ and $Prop(Q',d')$.

Encouraged by this observation, one is tempted to say boldly that the (closed linear) subspaces of $\mathcal{H}$ not only represent single quantum propositions but they also represent the logical relations between them in the sense that the set of all (closed, linear) subspaces considered together with the set theoretical operations $\cap$ (meet), $\cup$ (union) and $\setminus$ (set theoretic complement) also represent the quantum propositional system:

$$P^Q(d) \cup P^{Q'}(d')$$

representing

$$Prop(Q,d) \vee Prop(Q',d')$$

where

$Prop(Q,d) \vee Prop(Q',d') =$
"Q has value in d or Q' has value in d' with probability one"

and $\mathcal{H} \setminus P^Q(d)$ representing the negation of $Prop(Q,d)$.

The idea, in this form, is flawed: whereas the intersection of two closed linear subspaces is again a closed linear subspace, the union of two closed linear subspaces and the set theoretical complement of a closed linear subspace are not closed linear subspaces. However, the closure

$$P^Q(d) \vee P^{Q'}(d')$$

of the sum

$$P^Q(d) + P^{Q'}(d')$$

and the orthogonal complement

$$P(Q,d)^\perp$$

are again closed linear subspaces; so one is led to this question: Could perhaps the idea be salvaged nevertheless by this refinement: the set of all closed linear subspaces of $\mathcal{H}$ and its structure defined by the operations $\cap, \vee$ and $\perp$ represents the logic of the quantum propositions?

Quantum logic, in first approximation, is the discipline in which (mainly positive) answers to this question are worked out in detail.

The discipline is about sixty years old, its birth is commonly identified with the appearence of the 1936 paper of G. Birkhoff and J. von Neumann, [21]; however, as it will be seen in detail, the original Birkhoff-von Neumann idea of quantum logic is subtly different from what became later – and still is – the standard view. Explaining the original idea – both mathematically and in its historic context – is one of the central themes of this work. It should also be mentioned that the idea of considering the set of projections of the Hilbert space associated with a quantum system as the set of propositions regarding the values of the observables of the system is implicitly present already in von Neumann's ground-breaking 1927 paper [165] on the mathematical foundations of quantum mechanics (see especially the "Zusammenfassung" in [165]), and this idea got a rather systematic albeit short treatment already in von Neumann's 1932 book (section III. 5 in [168]).

Since the 1936 paper of Birkhoff and von Neumann quantum logic has become a vast, mixed field lying at the crossroads of and drawing on the methods of physics, mathematics, logic and philosophy. The approaches to and the interpretations of quantum logic have become so diverse in the past sixty years that it is impossible to cover even the main ideas of the most significant branches in a book of readable size. Consequently, despite the resulting unavoidable one-sidedness, choice must be made in what aspect of quantum logic and in what depth to cover. The present work was prepared on the basis of the following guiding/selecting principles:

INTRODUCTION

Aims:

- To give a concise introduction to what can be considered as the standard core of quantum logic.
- To make more accesible certain material, standard in the theory of operator algebras but much less known and inexcusably neglected in quantum logic textbooks, which is, however, indispensable to understand the original Birkhoff-von Neumann concept of quantum logic as expressed in [21].
- To call the attention to some specific areas, selected by the author's individual interest with the intention to formulate some open problems that can be subjects of future investigations.

Scope:

- Emphasis is put mainly on the description of the two mathematical structures (Hilbert lattice and von Neumann lattice) that form the foundation of quantum logic. No attempt is made to remain on the level of abstract orthomodular structures ("abstract quantum logic"). The basic facts of the structures described are given with detailed proofs, however.
- A standard argumentation is described in some detail which specifies – in terms of the logical notions of syntax-semantic – in what sense the Hilbert lattice (or the lattice of projections of a von Neumann algebra) can be considered as the logic of a quantum system described by a Hilbert space (respectively by the von Neumann algebra).
- The specific topics selected to be covered are the problem of hidden variables, the elementary theory of quantum conditionals, the problem of independence in quantum logic approach and the problem of existence of a statistical common cause of distant (superluminal) statistical correlations predicted by relativistic quantum field theory.

Structure/Overview:

Chapter 2 recalls the two main ingredients of the kinematical part of the Hilbert space formalism of quantum mechanics, namely the notions of observable and state. Also, the two key theorems in connection with the concepts of observable and state, the spectral theorem and Gleason theorem are spelled out. As a specific example of observables, the complementary observables (both the bounded and the unbounded ones) are described in some detail. After formulating the proposition known as Heisenberg uncertainty relation (or Heisenberg theorem), the notion of entropic uncertainty is defined, and an entropic uncertainty relation is proved in finite dimension. Both the complementary observables and the entropic uncertainty relation will feature in proofs in later chapters.

After a brief review in Chapter 3 of the elementary notions of lattice theory which are used in the book, the lattice of closed linear subspaces of a Hilbert space (the Hilbert lattice) is described in Chapter 4. In Section 4.1 the lattice operations between closed linear subspaces of a Hilbert space $\mathcal{H}$ and the resulting non-distributive, atomic lattice structure are defined, and propositions are proved showing that this lattice cannot be mapped into a Boolean lattice by certain homomorphisms. Special attention is paid to the differences of the Hilbert lattices of finite dimensional as opposed to infinite dimensional Hilbert spaces. In Section 4.2 the relation between the lattice theoretic and the algebraic operations, the latter ones defined for projections on $\mathcal{H}$, are summed up.

The atomic, non-distributive, non-modular, orthomodular Hilbert lattice of projections on an infinite dimensional HIlbert space is what is commonly called "concrete quantum logic", and the Chapter 5 outlines an interpretation of this concrete quantum logic as logic. Section 5.1 describes the general framework known as the "semantic approach" to physical theories, and in Section 5.2 the semantic approach is applied to classical mechanics to show that the logic, understood as the Tarski-Lindenbaum algebra of the propositional system defined by a classical mechanical system, is a Boolean algebra. Section 5.3 argues that something very similar holds for the Hilbert lattice: this lattice represents the equivalence classes (with respect to the equivalence relation "A is true if and only if B is true") of propositions A, B of the form "observable Q has its value in the set d of real numbers with probability one".

Chapter 6 is devoted to the theory of von Neumann algebras and von Neumann lattices. Section 6.1 gives the basic definitions in connection with C^*-and von Neumann algebras, Section 6.2 proves von Neumann's double commutant theorem and its consequence that the projections of a von Neumann algebra form an orthomodular, complete lattice. This section also describes the Murray-von Neumann dimension theory of projections, which is intimately related to the classification theory of factors. The classification theory of factors also is recalled briefly in this section. The main purpose of Section 6.2 is to describe the lattices of finite von Neumann algebras, in particular the projection lattice of a type $\mathbf{II}_1$ factor, which is modular and non-atomic – a surprise. The von Neumann lattice of projections of a type $\mathbf{II}_1$ factor von Neumann algebra is what Birkhoff and von Neumann considered "quantum logic" – the reason why is the topic of Chapter 7. Chapter 7 not only recalls the Birkhoff-von Neumann concept of quantum logic but also tries to put their view in perspective by explaining the conceptual and historical background of their (especially von Neumann's) view. It is seen in Chapter 7 that the Birkhoff-von Neumann concept, and in particular von Neumann's preference of the type II_1 algebra, is related to why von Neumann lost his belief in the Hilbert space formalism of quantum mechanics by the time 1936. This is not widely known, and in any case it is a neglected theme in both quantum logic and in the history of quantum mechanics; thus the material in Chapter 7 has some historical interest in itself.

Section 8.1 in Chapter 8 describes the elementary theory of the quantum conditional connective. The basic properties of the three quantum conditional connectives that can be written as lattice polinoms and which satisfy the minimal implicative criteria are presented. In particular, classical implicative criteria are listed that the quantum conditionals violate, and it is shown that one of the conditionals, the so-called Mittelstaedt conditional (also known as "Sasaki hook") is a counterfactual conditional in the sense of Stalnaker's possible world semantics, if the set of possible worlds is identified with the elements of the Hilbert space and the similarity of the possible worlds is measured in Hilbert space norm. The subject of Section 8.2 is the problem of statistical inference and the related issue of conditionalization. Having recalled the theory of conditionalization and in particluar the notion of conditional expectation in classical probability theory, and having formulated the problem of statistical inference in terms of von Neumann algebras, Section 8.3 shows that the so-called Stalnaker's Thesis (="probability of conditional=conditional probability") breaks down in quantum logic, i.e. that the probability of the quantum conditional is not equal to the quantum conditional probability, if the latter one is given by a non-commutative conditional expectation.

Chapter 9 is devoted to the hidden variable problem. After a brief (hence simplifying) and non-technical review in Section 9.1 of some episodes in the long and confusing history of the problem, a definition of hidden variable theory of quantum mechanics is given in the operator algebraic framework of quantum mechanics. On the basis of that definition the hidden variable problem is re-interpreted as the task of determining the algebraic structures in the set of observables that cannot be preserved under reduction of the statistical character of quantum mechanics. The statistical content (uncertainty) of a probabilistic physical theory can be measured either by the pointwise dispersion of states in the theory, or by an appropriately defined entropy, and the corresponding "dispersive" and "entropic" hidden variable theories are considered in Sections 9.2 and 9.3. Propositions are presented in Section 9.2 and 9.3 showing that the Jordan algebra structure of the observables is rigid i.e. it cannot be preserved under reduction of statistical uncertainty (either measured by dispersion or by entropy), and it also is proved that the algebra of observables cannot be embedded into a larger one so that the physical theory determined by the larger algebra of observables be less statistical – if the uncertainty is measured by dispersion. Whether the entropic version of this statement is also true in general, remains an open problem. It is proved in Section 9.3, however, that such an embedding does not exist in the entropic case either – provided the algebras involved are finite dimensional, complete matrix algebras. Section 9.4 is devoted to the analysis of the consisteny problem formulated first by Bell: the problem of whether reduction of the statistical uncertainty of quantum mechanics by a hidden variable theory is compatible with preserving physical locality. Bell's treatment of the problem is summarized first, and the notion of local hidden variable theory is given in Section 9.4 in terms of quasilocal algebras of relativistic quantum field theory. A proposition is proved then that isolates relativistic locality properties that cannot be preserved under the reduction of the statistical character of quantum field theory.

In the first section of Chapter 10 the main idea and the basic ingredients of the theory of local, algebraic relativistic quantum field theory are summarized together with a few important theorems that characterize this theory, or which are referred to at some places in the book. Based on this brief review, Section 10.2 introduces the notion of Bell correlation in operator algebraic terms, isolates conditions implying that the Bell correlation is bounded by 1 (which is Bell's inequality) and presents propositions spelling out the violation of Bell's inequality for observables in local observable algebras pertaining to spacelike separated spacetime regions. The upshot of the violation of Bell's inequality in relativistic quantum field theory, pointed out at the end of Section 10.2, is that local relativistic quantum field theory predicts superluminal statistical correlations, i.e. statistical correlations between events that are spacelike separated. Section 10.4 formulates the two strategies that one can follow in principle to explain correlations: one is based on assuming a direct causal influence between the correlated events, the other one assumes a common cause of the correlation.

The existence of direct causal influence between the correlated events raises the problem of whether the event structures (logics) involved are independent of each other. So the problem of superluminal correlations leads to the problem of independence of two subsystems of a quantum system. This problem is the subject of Chapter 11. Section 11.1 raises the problem of logical independence of two sub-quantum logic of a quantum logic, where "quantum logic" means von Neumann lattice. The definition of logical independence proposed expresses the semantic independence of sublattices and, after motivating this definition, several propositions are proved which relate logical independence to other (statistical) independence conditions. Section 11.2 specifies the idea (put forward by David Lewis) of counterfactual probabilistic independence in terms of the von Neumann lattices that appear naturally in algebraic relativistic quantum field theory as representatives of events that are localized in regions in the Minkowski spacetime. It is shown in particular that under a certain specification in quantum field theory of Stalnaker's possible world semantics of the truth values of counterfactuals, the violation of the Bell's inequality in relativistic quantum field theory does not imply that spacelike separated events are not independent in the probabilistic counterfactual sense. This chapter contains explicitly formulated open problems in connection with the characterization of logically independent lattices, and there remain open questions also regarding the counterfactual probabilistic independence.

Since the results on the independence of local observable algebras in relativistic quantum field theory indicate that the algebras are independent in spite of the presence of statistical correlation between events in them, one is led to the question of whether there exist common causes of the superluminal correlations. Chapter 12 is devoted to the analysis of this problem. Section 12.1 recalls Reichenbach's notion of a probabilistic common cause of statistical correlation and, after pointing out a couple of open questions concerning this notion, Section 12.2 specifies Reichanbach's common cause principle in terms of algebraic relativistic quantum field theory. The problem is raised then, whether quantum field theory satisfies the common cause principle, i.e. whether the superluminal correlations have a probabilistic common cause in Reichenbach's sense. This remains an open problem. All that can be, and is in fact, shown in Section 12.2 is that existence of genuinely probabilistic common causes of all superluminal correlations predicted by the vacuum state implies the statistical independence of the local algebras.

The bibliography was made extensive enough to be of help if the reader decides to seriously study quantum logic beyond what the present work can offer either in scope or in depth. No attempt has been made to present a complete bibliography on quantum logic, however. (There exists a comprehensive bibliography on quantum logic, see the Bibliographic notes to this Introduction.) To help the reader and to facilitate further study, each chapter is closed with a section entitled "Bibliographic notes" that locate the primary source of the results presented, and give references to works closely related to the topic of the chapter.

Assumptions:

The reader is assumed to be familiar with the elements of functional analysis, in particular with the theory of Hilbert spaces. Knowledge of classical probability theory (in measure theoretic form) and familiarity with the distinction syntax-semantic and the associated logical notions is helpful. As a rule, the later chapters assume more, and the Chapters 6 and 11, especially the Section 6.3, while largely selfcontained, are more demanding technically. Although not necessary to understand the presented material formally, the whole issue of quantum logic can probably be appreciated only on the basis of knowing the elementary Hilbert space quantum mechanics.

My intention was to expand the original lecture notes so that the book also can be used to teach quantum logic. Skipping the more technical sections (especially those on the von Neumann algebra theory such as 6.3 in Chapter 6, and leaving out the proofs in particular), the book should be usable as a core text in a one semester introductory course to quantum logic on the upper undergraduate level. Together with the operator algebraic part the book should be suitable for a graduate course, and discussing the loose ends and open problems in the Chapters 11 and 12 can be used to introduce students to some of the questions that are being currently debated in the foundational literature. The ideal reader I had in mind while writing was a somewhat philosophically minded physicist with a strong respect for and interest in mathematics.

1.1. Bibliographic notes

While there are several books on quantum logic available (e.g. [164] [112] [18] [113] [116]), I know of no work that presents quantum logic mainly in the von Neumann algebra framework, or which contains a substantive discussion of quantum logic and the related issues from the perspective of von Neumann lattices. Many books, especially those dealing with foundational questions of quantum mechanics also devote at least a few pages to the issue of quantum logic; let us mention [134] in this category. Varadarajan [164] concentrates on the mathematical aspects, [18] contains an especially useful review of the representation theory of abstract orthomodular lattices, itself a vast, rather mathematical subfield of quantum logic, to which C. Piron made significant contributions, and which is not even mentioned here. Piron, too, wrote his own book [112]. Though Jauch considers calling $\mathcal{P}(\mathcal{H})$ "logic" a mistake (he refers to $\mathcal{P}(\mathcal{H})$ as the "quantum propositional system", the Chapter 5 in his book [75] has become a classic reference in quantum logic. Quantum logic as the theory of abstract orthomodular lattices is treated in [116]. An interesting recent book offering some nonstandard interpretation of quantum logic is Pitowsky's work [113]. Many of the basic papers on quantum logic can be found in the two volume collection edited by C.A. Hooker. The first volume, [70], contains the classic papers, both the technical and the more philosophic ones (e.g. it contains the Birkhoff-Neumann paper [21]), whereas the second volume [71] makes the current (at the time of the publication) papers easily accessible. There exists a comprehensive bibliography on quantum logic compiled by Pavicic [108].

CHAPTER 2

Observables and states in the Hilbert space formalism of quantum mechanics

In this chapter the main elements of the Hilbert space formalism of quantum mechanics are summed up briefly. Usually this formalism is considered *the* formalism of quantum mechnics but it is not the only and not even the most "natural" one. We shall see in Chapter 6 that the so-called "algebraic" quantum mechanics is equally important. The Hilbert space formalism has a long history that goes back to the twenties and it was worked out rigorously by von Neumann in [168]. In Chapter 7 we shall have to say a few words about von Neumann's contribution to establishing this formalism systematically and about his subsequent critical attitude towards the Hilbert sapce formalism. In the present chapter only the two key concepts, the observables and states and their mathematical representatives are discussed. Summarizing the formalism also serves the purpose of fixing some of the notations that will be used subsequently. As far as possible, the formalism is presented here without any interpretation, that is, without addressing and discussing those controversial interpretative issues that have been with us since the birth of this theory. One of these problems is whether what is called "quantum logic" can be viewed as the logic of the quantum system. This issue will be discussed in Section 5.3.

2.1. Observables

The basic concepts of the Hilbert space quantum mechanics are the *state* of the system, the *observable quantities (observables)*, the *possible values* of the observables and the *expectation values* of the observables.

The system is conceived as being in a definite state at every moment. The observables are considered as being determined independently of the state of the system. Unlike the physical quantities in classical mechanics the quantum observables can not take on arbitrary values: the set of possible values of every observable is determined by the special features of the system. The preparation of the system in a state does not determine which possible values the observables take in that state, fixing a particular state determines uniquely the expectation values of the observables, however.

The mathematical representatives of the basic concepts is carried by a (typically) infinite dimensional, complex, separable Hilbert space $\mathcal{H}$. Recall that the complex linear space $\mathcal{H}$ is a Hilbert space if there is a scalar product $\langle \eta, \zeta \rangle$ defined between the elements $\eta, \zeta \in \mathcal{H}$ such that $\mathcal{H}$ is complete with respect to the norm $\| \eta \|$ defined by

$$\| \eta \|^2 \equiv \langle \eta, \eta \rangle \tag{2.1}$$

(That is to say, every Cauchy sequence in norm has a limit in $\mathcal{H}$.)

$\mathcal{H}$ is said to be *separable* if there exists a countable dense set in $\mathcal{H}$. The elements $\eta_n \in \mathcal{H}$, $n = 1, \ldots N$ are *linearly independent* if $\sum_n^N \lambda_n \eta_n = 0$ implies $\lambda_n = 0$ for all $n = 1, \ldots, N$. The Hilbert space $\mathcal{H}$ is called n *dimensional*, $\dim(\mathcal{H}) = n$, if $\mathcal{H}$ is n dimensional as a linear space, i.e. if the maximal number of linearly independent elements in $\mathcal{H}$ is equal to n. $\mathcal{H}$ is said to be *finite dimensional* if there is an n, such that $\dim(\mathcal{H}) = n$. If $\mathcal{H}$ is not n dimensional for any n, then $\mathcal{H}$ is said to be *infinite dimensional*.

The (unique up to isomorphism) examples of separable Hilbert spaces are the square integrable complex functions $L^2(\mathbb{R}, \mu)$ defined on $\mathbb{R}$, and the set of square-summable series $\ell^2(\mathbb{N})$.

The observable quantities of the system are represented by (generally unbounded) selfadjoint operators Q defined on $\mathcal{H}$. The spectrum $\sigma(Q)$ of Q is the "set of possible values of Q". Recall that

$$\sigma(Q) = \mathbb{C} \setminus R(Q)$$

where $R(Q)$ is the resolvent set of Q, which is the set of numbers $q \in C$ for which the operator $(Q - qI)$ has a bounded inverse in $\mathcal{B}(\mathcal{H})$ (here I is the identity operator on $\mathcal{H}$).

If Q is selfadjoint, then its spectrum is a subset of the real numbers. The number q is an eigenvalue of Q if there is a non-zero $\eta \in \mathcal{H}$ such that $Q\eta = q\eta$. If q is an eigenvalue of Q, then $q \in \sigma(Q)$. The converse is not true: the spectrum can contain elements that are not eigenvalues.

Example: Let Q be the position operator on $L^2(\mathbb{R}, \mu)$: The operator Q is defined on the domain

$$D(Q) \equiv \{f \in \mathcal{H}|\ id_{\mathbb{R}} f \in \mathcal{H}\}$$

($id_{\mathbb{R}}$ denoting the identity map on $\mathbb{R}$: $id_{\mathbb{R}}(x) = x$, $x \in \mathbb{R}$) by

$$(Qf)(x) = xf(x)\ (Qf = id_{\mathbb{R}}f)\ (f \in D(Q)) \tag{2.2}$$

Q is densely defined since $D(Q)$ contains the square integrable functions with compact support, which are dense in $L^2(\mathbb{R}, \mu)$. The spectrum of Q is the whole real line, and Q has no eigenvalues. Q is not bounded since the functions

$$f_n(x) = \begin{cases} 1, & \text{if } x \in [n, n+1]; \\ 0, & \text{otherwise.} \end{cases}$$

are contained in the domain of Q and we have

$$\| f_n \| = 1$$

but

$$\| Q f_n \| \geq n$$

Q is not everywhere defined since the function below is not in the domain of definition

$$f^*(x) = \begin{cases} 1/x, & \text{if } x \geq 1; \\ 0, & \text{if } x < 1. \end{cases}$$

because the function

$$x \mapsto (Qf^*)(x) = x(1/x) = 1, \ (x \geq 1)$$

is not square integrable.

Q does not have eigenvalues: if q were an eigenvalue then there would exist a non-zero $f \in L^2(\mathbb{R}, \mu)$ such that

$$0 = \| Qf - qf \|^2 = \int f^2 \, |\, \mathrm{id}_\mathbb{R} - q\,|^2 \, d\mu$$

and so $f = 0$ μ-almost everywhere (here we used the spectral theorem, see below), thus $f = 0$.

Example: Let P be the momentum operator given on $L^2(\mathbb{R}, \mu)$:

$$(Pf)(x) = -i((d/dx)f)(x) \quad (Pf = -if')$$

This definition makes sense for all continuously differentiable functions with compact support, and these are dense in $L^2(\mathbb{R}, \mu)$. Thus P is densely defined. P is not bounded since for the functions f_n defined by

$$f_n = \begin{cases} \sin nx, & \text{if } x \in (-\pi, \pi); \\ 0, & \text{if } x \notin (-\pi, \pi) \end{cases}$$

it holds that

$$\begin{aligned} \| f_n \| &= \sqrt{2\pi} & (n = 1, 2, \ldots) \\ \| P f_n \| &= n\sqrt{2\pi} & (n = 1, 2, \ldots) \end{aligned}$$

P and Q are Fourier transform related:

$$P = \mathcal{F}^{-1} Q \mathcal{F}$$

where $\mathcal{F}$ is the unitary Fourier operator on $L^2(\mathbb{R}, \mu)$. Thus the spectrum of P is also the whole real line. P has no eigenvalues.

Furthermore, the "Heisenberg commutation rule" holds for Q and P

$$(QP - PQ)f = if \tag{2.3}$$

for all f for which the left hand side of (2.3) is meaningful. Q and P are the position and momentum operators of the non-relativistic free particle in one degree of freedom.

A distinguished role play the observables that have just two possible "values": these are represented by projections:

Definition 2.1 The operator A is a *projection* if it is selfadjoint and idempotent, i.e. $A^2 = A = A^*$.

The projections are bounded operators, if $A \neq 0$ and A is a projection then the norm of A is equal to one. The set of all projections is denoted by $\mathcal{P}(\mathcal{H})$. Special projections are the zero 0 and the identity operator I.

Definition 2.2 Two projections A, B are called *orthogonal* (or *disjoint*) if $AB = BA = 0$.

The distinguished status of the projections is reflected by the fact that all selfadjoint operators can be obtained from them. This is the content of the important spectral theorem. Before formulating the spectral theorem let us recall the notion of a projection valued measure.

Definition 2.3 The map

$$\mathrm{P}: \mathcal{B}(\mathbb{R}) \to \mathcal{P}(\mathcal{H})$$

is called a *projection valued measure* if it has the following properties:

(i) $\mathrm{P}(0) = 0$
(ii) $\mathrm{P}(I) = I$
(iii) $\mathrm{P}(\cup_i d_i) = \vee_i \mathrm{P}(d_i)$ for pairwise orthogonal real Borel sets $d_i \in \mathcal{B}(R)$, in which case $\vee_i \mathrm{P}(d_i) = \sum_i \mathrm{P}(d_i)$
(iv) $\mathrm{P}(\mathbb{R} \setminus d) = I - \mathrm{P}(d) = \mathrm{P}(d)^\perp$

If P is a projection valued measure, then the measure $\mu^\mathrm{P}(\zeta, \eta)$ defined by

$$d \mapsto \mu^\mathrm{P}(\zeta, \eta)(d) = \langle \zeta, \mathrm{P}(d) \eta \rangle \tag{2.4}$$

is a complex measure.

Proposition 2.1 (Spectral theorem for selfadjoint operators) *If Q is a selfadjoint operator then there is a unique projection valued measure P such that*

$$\langle \zeta, Q\eta \rangle = \int id_{\mathbb{R}} d\mu^{\mathrm{P}}(\zeta, \eta) \quad \eta \in D(Q), \zeta \in \mathcal{H} \tag{2.5}$$

and, conversely, every projection valued measure P determines a normal operator Q through (2.5) (Q is normal if $D(Q) = D(Q^*)$ and $QQ^* = Q^*Q$).

Since the the projection valued measure P depends on Q, this will be indicated by using the notation P^Q if necessary. P^Q is also called "the spectral measure of Q". This terminology is justified by the following fact: The support supp(P) of a projection valued measure P is, by definition, the smallest closed set for which $\mathrm{P}(\mathrm{supp}(\mathrm{P})) = I$. It holds that

$$\sigma(Q) = \mathrm{supp}(\mathrm{P}^Q)$$

i.e. the spectrum of the operator is equal to the support of the projection valued measure determined by the operator through the spectral theorem.

One can use the spectral theorem to define functions of selfadjoint operators. If Q is a selfadjoint operator, P is its spectral measure and $f: \mathbb{R} \to C$ is a Borel measurable function, then the set

$$D_f \equiv \{\xi \in \mathcal{H} \mid \int |f|^2 d\mu^{\mathrm{P}}(\xi, \xi) \leq \infty\} \tag{2.6}$$

is a linear, dense subspace in $\mathcal{H}$. One then defines $f(Q)$ as the linear operator that has D_f as its domain, and for which

$$\langle f(Q)\xi, \eta \rangle \equiv \int f d\mu(\xi, \eta) \quad \xi \in D_f, \eta \in \mathcal{H}$$

Example: If the spectrum of Q contains the q_i eigenvalues only and E_i denotes the projection belonging to the i-th eigenvalue, then $Q = \sum_i q_i E_i$. In particular, if Q is a projection, $Q = E$, then its spectral resolution is given by

$$Q = 1E + 0(I - E) = 1E + 0E^{\perp}$$

Example: The spectral measure of the position operator Q is given by multiplication by the characteristic (indicator) function χ:

$$(\mathrm{P}^Q(d))f(x) = \chi_d(x)f(x) \quad (f \in L^2(\mathbb{R}, \mu));$$

and the spectral measure of the momentum operator P is given by

$$(\mathrm{P}^P(d))f(k) = \int_d \exp(-ikx)f(x)dx$$

Example: A special example of observables is the *complementary* pair of observables. These observables will play a role in the proof of Proposition 9.11 so we describe them here in some detail.

Definition 2.4 Let A and B be two selfadjoint operators on the complex Hilbert space $\mathcal{H}_n$ of dimension $2 < n < \infty$; i.e. A and B are complex n by n matrices such that all their eigenvalues are real. Let η_j^A, η_k^B ($j,k = 1, 2, \ldots n$) denote the eigenvectors of A and B respectively. A and B are said to be *complementary* to each other (or the pair (A, B) is said to be a *complementary pair*) if none of their eigenvalues is degenerate and the following holds:

$$|\langle \eta_j^A, \eta_k^B \rangle| = \frac{1}{\sqrt{n}} \quad \text{for all} \quad j, k = 1, 2, \ldots n \tag{2.7}$$

One can write (2.7) in the equivalent form

$$\langle \eta_j^A, \eta_k^B \rangle = \frac{1}{\sqrt{n}} e^{ix_{jk}} \tag{2.8}$$

where $e^{ix_{jk}}$ are arbitrary phase factors.

By explicitly constructing complementary operator pairs in each finite dimension $n \geq 2$ we show next that there exist complementary operators indeed.

Let η_j^A ($j = 1, 2 \ldots n$) be an arbitrary orthogonal system of unit vectors in $\mathcal{H}_n$, and let c_j ($j = 1, 2, \ldots n$) be the nth roots i.e.

$$c_j = e^{j(2\pi i/n)}$$

If η_k^B is defined by

$$\eta_k^B = \frac{1}{\sqrt{n}} \sum_j (c_k)^j \eta_j^A \quad (k = 1, 2, \ldots n) \tag{2.9}$$

then we caim that the operators A, B defined by η_j^A and η_k^B as their eigenvectors are complementary. To see this, one only has to show that η_k^B is an orthogonal system in $\mathcal{H}_n$ because (2.7) holds trivially.

$$\langle \eta_k^B, \eta_m^B \rangle =$$
$$\frac{1}{n} \langle \sum_j (c_k)^j \eta_j^A, \sum_j (c_m)^j \eta_j^A \rangle = \frac{1}{n} \sum_j (c_m^*)^j (c_k)^j =$$
$$\frac{1}{n} \sum_j e^{(m-k)j(2\pi i/n)}$$

It is obvious that $\sum_j e^{(m-k)j(2\pi i/n)} = 1$ if $m = k$; furthermore, if $m \neq k$, then to show $\sum_j e^{(m-k)j(2\pi i/n)} = 0$ one has to see that $\sum_j (c_m)^j = 0$ if $c_m \neq 1$ is any of the nth roots. This, however, is clear because $\sum_j (c_m)^j = \sum_j c_j = 0$.

Next we show that the complementary pair constructed above satisfies the discrete analog of the Weyl form of the canonical commutation relation.

Proposition 2.2 *Let A, B be the pair of complementary operators just constructed. Then we have*

$$e^{ijA} e^{imB} = e^{jm(-2\pi i/n)} e^{imB} e^{ijA} \qquad (2.10)$$

Proof: One can express η_k^A from (2.9) in terms of η_j^B:

$$\eta_k^A = \frac{1}{\sqrt{n}} \sum_j (c_j^*)^k \eta_j^B \qquad (2.11)$$

Now since the eigenvalues do not enter the definition of complementarity, they may be chosen arbitrarily as

$$A\eta_j^A = (2\pi j/n)\eta_j^A \qquad B\eta_k^B = (2\pi k/n)\eta_k^B \qquad (2.12)$$

With the help of (2.9) and (2.12) we can compute $e^{ijA}\eta_k^B$ as follows

$$e^{ijA}\eta_k^B =$$

$$e^{ijA}\frac{1}{\sqrt{n}}\sum_m (c_k)^m \eta_m^A = \frac{1}{\sqrt{n}}\sum_m (c_k)^m e^{ijA}\eta_m^A$$

$$= \frac{1}{\sqrt{n}}\sum_m (c_k)^m e^{jm(2\pi i/n)}\eta_m^A = \frac{1}{\sqrt{n}}\sum_m e^{km(2\pi i/n)} e^{jm(2\pi i/n)}\eta_m^A$$

$$= \frac{1}{\sqrt{n}}\sum_m e^{(k+j)m(2\pi i/n)}\eta_m^A$$

$$= \begin{cases} \eta_{k+j}^B, & \text{if } k+j \leq n; \\ \eta_{k+j-n}^B, & \text{if } k+j > n \end{cases}.$$

or, in short

$$e^{ijA}\eta_k^B = \eta_{[k+j]}^B \qquad (2.13)$$

where $[k+j] = k+j$ modulo n.

With the help of (2.12) and (2.13) we can do the following easy computation

$$e^{ijA}\exp(imB)\eta_k^B = e^{ijA}\exp(mk(2\pi i/n))\eta_k^B$$

$$= e^{mk(2\pi i/n)}\eta_{[k+j]} \tag{2.14}$$

$$e^{jm(-2\pi i/n)}e^{imB}e^{ijA}\eta_k^B = e^{jm(-2\pi i/n)}e^{imB}\eta_{[k+j]}^B \tag{2.15}$$

$$= e^{jm(-2\pi i/n)}e^{m[k+j](2\pi i/n)}\eta_{[k+j]}^B \tag{2.16}$$

$$= e^{m([k+j]-j)(2\pi i/n)}\eta_{k+j}^B \tag{2.17}$$

If $k+j \leq n$ then $[k+j]-j = k$, thus in this case the right hand sides of (2.14) and (2.15) are the same, if, furthermore, $k+j > n$, then $[k+j]-j = k-n$ and then

$$e^{m(k-n)(2\pi i/n)} = e^{mk(2\pi i/n)}[e^{-2\pi i}]^m$$
$$= e^{mk(2\pi i/n)}$$

Thus the right hand sides, and, consequently, the left hand sides of (2.14) and (2.15) are the same for all k, and this means that the matrix equality (2.10) holds. □

Since (2.10) is but the discrete analog of the Weyl form of canonical commutation relation, the complementary operators are called *canonical complementary operators*.

An important feature of a complementary observable pair in finite dimension is that they satisfy an entropic uncertainty relation maximally (see Proposition 2.9).

The definition of complementarity (Definition 2.7) makes sense only in finite dimension. It is natural to ask whether there is any reasonable generalization of the complementarity definition which would allow possibly unbounded selfadjoint operators defined in an infinite dimensional complex Hilbert space to be complementary? The answer is yes. To give this general definition of complementarity first we reformulate the (2.7) in terms of the spectral measures of the observables A and B.

Let $p(j) = \frac{1}{n}$ ($j = 1, 2, \ldots n$) be the uniform probability measure on $\hat{N} = \{1, 2, \ldots n\}$, and let P^A, P^B be the spectral measures of two selfadjoint operators defined on $\mathcal{H}_n$ with non-degenerate spectrum. We claim that (2.7) is equivalent to the following condition

$$\mathrm{Tr}(P^A(I)P^B(J)) = p(I)p(J) \quad \text{for all } I, J \subset \hat{N} \tag{2.18}$$

where Tr is the normalized trace over $\mathcal{B}(\mathcal{H}_n)$. To see this, one just has to reformulate (2.18) in the following way

$$Tr(P^A(I)P^B(J)) =$$

$$\frac{1}{n}\sum_{k}\langle\eta_k^B,\sum_{j\in I}\mathrm{P}^A(j)\sum_{m\in J}\mathrm{P}^B(m)\eta_k^B\rangle = \frac{1}{n}\sum_{k\in J}\langle\eta_k^B,\sum_{j\in I}\mathrm{P}^A(j)\eta_k^B\rangle =$$

$$\frac{1}{n}\sum_{k\in J}\langle\eta_k^B,\sum_{j\in I}\mathrm{P}^A(j)\sum_{m}\langle\eta_k^B,\eta_m^A\rangle\eta_m^A\rangle =$$

$$\frac{1}{n}\sum_{k\in J}\langle\eta_k^B,\sum_{m}\langle\eta_k^B,\eta_m^A\rangle\sum_{j\in I}P_j^A\eta_m^A\rangle = \frac{1}{n}\sum_{k\in J}\sum_{m\in I}\langle\eta_k^B,\langle\eta_k^B,\eta_m^A\rangle\eta_m^A\rangle =$$

$$\frac{1}{n}\sum_{k\in J}\sum_{m\in I}\langle\eta_k^B,\eta_m^A\rangle^*\langle\eta_k^B,\eta_m^A\rangle = \frac{1}{n}\sum_{k\in J}\sum_{m\in I}|\langle\eta_k^B,\eta_m^A\rangle|^2$$

If (2.7) holds then

$$\frac{1}{n}\sum_{k\in J}\sum_{l\in I}|\langle\eta_k^B,\eta_l^A\rangle|^2 = \frac{1}{n}\#(J)\times\#(I)\times\frac{1}{n} = p(J)p(I).$$

Conversely, one gets (2.7) from (2.18) by chosing $J=\{j\}, I=\{k\}$

The definition (2.18) of complementarity is suitable for generalization to the infinite dimensional case:

Definition 2.5 Two selfadjoint operators A, B on the infinite dimensional Hilbert space $\mathcal{H}$ are called *complementary* to each other if

$$Tr(\mathrm{P}^A(E)\mathrm{P}^B(G)) = (1/2\pi)\mu(E)\mu(G)$$

for any two Borel sets E, G such that $\mu(E), \mu(G) < \infty$, where μ is the Lebesgue measure on the set of real numbers.

It is clear that if A, B are complementary in the sense of the above definition, then they cannot be bounded.

Proposition 2.3 *Let Q and P be the canonically conjugate observables of position and momentum of the free particle moving in one dimension. Then Q and P are complementary to each other.*

Proof: Let $\mathcal{F}$ and $\mathcal{F}^{-1}$ be the Fourier operator and its inverse on $L^2(\mathbb{R},\mu)$. The spectral projector $\mathrm{P}^Q(E)$ is the multiplication by the characteristic (indicator) function χ_E:

$$\mathrm{P}^Q(E)f = \chi_E f \qquad (f\in L^2(\mathbb{R},\mu))$$

consequently,

$$\mathrm{P}^Q(E)\mathrm{P}^P(G)f = \chi_E \mathcal{F}^{-1}\chi_G\mathcal{F}f$$

for $f\in L^2(\mathbb{R},\mu)$. Since $\mu(E), \mu(G)$ are finite, the functions χ_E, χ_G are square integrable, thus the operator $\mathrm{P}^Q(E)\mathrm{P}^P(G)$ is an intergral operator with kernel

$$K(x,y) = \chi_E(x)(\mathcal{F}^{-1}\chi_G)(x-y)$$

that is we have

$$(P^Q(E)P^P(G))(f)(x) = \int \chi_E(x)(\mathcal{F}^{-1}\chi_G)(x-y)f(y)d\mu(y)$$

This means that $P^Q(E)P^P(G)$ is a Hilbert-Schmidt operator with the Hilbert-Schmidt norm

$$\| P^Q(E)P^P(G) \|_{Tr} =$$
$$\int | K(x,y) |^2 d\mu(x)d\mu(y) =$$
$$\int | \chi_E(x) |^2 | (\mathcal{F}^1\chi_G)(x-y) |^2 d\mu(x)d\mu(y) = (1/2\pi)\mu(E)\mu(G)$$

On can write

$$\| P^Q(E)P^P(G) \|_{Tr} =$$
$$Tr(P^Q(E)P^P(G)(P^Q(E)P^P(G))^*) = Tr(P^Q(E)P^P(G)P^Q(E))$$
$$= Tr(P^Q(E)P^P(G))$$

In the last equality we used the fact that $P^Q(E)P^P(G)$ is a trace class operator. □

Note that the following problem is open

Problem 2.1 Do there exist unbounded complementary operator pairs that are not unitary equivalent to the canonical complementary pair Q, P?

We close this section by mentioning another characteristic property of the complementary operators Q and P. If d is a bounded real interval, and $\xi \in \mathcal{H}$ is a state vector such that

$$\langle P^Q(d) \rangle_\xi = 1$$

i.e. the probability of finding the particle in the set d is equal to 1, then the support of ξ is in d, hence by the Paley-Wiener Theorem the support of the Fourier transform of ξ is the whole real line. Thus the probability that the value of P lies in any bounded interval s cannot be equal to 1. This means that the canonical operator pair Q and P also have the following property, which sometimes also is referred to as "complementarity":

$$P^Q(d) \wedge P^P(s) = 0 \quad \text{for all bounded real intervals} \quad s, d \qquad (2.19)$$

2.2. States

Definition 2.6 The mathematical representatives of the physical states are the

$$\varphi: \mathcal{P}(\mathcal{H}) \to [0,1]$$

maps that have the following (i)-(iii) properties:

OBSERVABLES AND STATES

(i) $\varphi(0) = 0$
(ii) $\varphi(E^\perp) = 1 - \varphi(E)$
(iii) $\varphi(\vee_i E_i) = \sum_i \varphi(E_i)$ for mutually orthogonal E_i.

That is to say, a state φ possesses properties analogous to those of a classical probability measure; this is why they are often referred to as "non-commutative probability measures". An observable Q, which can be identified with a projection valued measure P^Q, together with a state φ determine a classical probability measure μ on the real line through $\mu = \varphi \circ P^Q$. Unlike classical statistical mechanics, where one usually deals with just one single probability measure, in quantum mechanics one has thus to deal with a whole bunch of classical probability measures.

Note that the existence of such non-commutative probability measures is not quite obvious; however, the next theorem shows that there are such measures.

Proposition 2.4 (Gleason's Theorem) *If $dim(\mathcal{H}) \geq 3$, then for every φ there exists a positive, trace-class operator w with trace equal to 1 such that $\varphi(E) = \mathrm{Tr}(wE)$, and, conversely, every positive trace-class operator w that has trace 1 determines a state φ through the formula $\varphi(E) = \mathrm{Tr}(wE)$.*

The trace class operator w in the Gleason's Theorem is called the *density matrix* (or density operator) determined by φ. Every density matrix has the form

$$w = \sum_i r_i P_{\xi_i} \tag{2.20}$$

with P_{ξ_i} being one dimensional projection operators spanned by the vectors ξ_i ($\sum_i r_i = 1$, $r_i \geq 0$) If the sum in (2.20) is finite, then w is called *finite range* density matrix. Typical states are the vector states that are determined by a single vector η in $\mathcal{H}$, in which case

$$\mathrm{Tr}(wE) = \| E\eta \|^2$$

The significance of Gleason's theorem is not only that it shows that there exist a lot of non-commutaticve probability measures. Equaly important is the fact that the theorem shows at the same time that the non-commutative probability measures can be extended from the lattice of projections to the set of bounded observables on the Hilbert space. This is because, given a w density operator, the expression

$$\phi(Q) = \mathrm{Tr}(Qw)$$

does make sense for any bounded Q operator on $\mathcal{H}$. That is to say, Gleason's theorem is a theorem of non-commutative integration.

The m-th *momentum* of the observable Q in the state φ is defined to be the number
$$M^m(w,Q) = \int (id_{\mathbb{R}})^m d\mu(w,P^Q) \tag{2.21}$$
if it exists. The measure $\mu(w,P^Q)$ is defined by
$$\mu(w,P^Q)(d) = \text{Tr}(P^Q(d)w)$$
If $m=1$ then
$$\langle Q \rangle_w \equiv M(w,Q) = \int id_{\mathbb{R}} d\mu(w,P^Q) \tag{2.22}$$
is called the *expectation value* of Q in the state w. If Q is a bounded operator, then
$$M^m(w,Q) = \text{Tr}(Q^m w)$$
and this is meaningful for every w. If Q is not bounded, then $M^m(w,Q)$ exists as a finite quantity if w is a finite range density matrix with P_i spanned by ξ_i and $\xi_i \in D(Q)$, and in this case
$$M^m(w,Q) = \sum_i r_i \langle \eta, Q^m \eta \rangle$$

The *dispersion* $\sigma_\phi(Q)$ of the state ϕ on the observable Q is defined by
$$\sigma_\phi(Q) \equiv \langle Q^2 \rangle_w - (\langle Q^2 \rangle_w)^2 \tag{2.23}$$
$$= \int id_{\mathbb{R}}^2 d\mu(w,P^Q) - \left(\int id_{\mathbb{R}} d\mu(w,P^Q)\right)^2 \tag{2.24}$$
if the left hand side exists as a finite quantity. If Q is a bounded observable, then
$$\sigma_\phi(Q) = \phi(Q^2) - \phi(Q)^2 \tag{2.25}$$
is finite and $\sigma_\phi(Q) \geq 0$. If $\sigma_\phi(Q) = 0$ for all selfadjoint Q, then the state ϕ is called *dispersion-free*. It is a characteristic property of the Hilbert space formalism that dispersion-free states do not exist. This fact is so important that we spell it out in the form of a proposition.

Proposition 2.5 (Von Neumann 1932) *If $\mathcal{H}$ is a complex Hilbert space of dimension greater than 2, then there exists no dispersion-free state on $\mathcal{H}$.*

Another characteristic feature of Hilbert space quantum mechanics is the existence of uncertainty relations. There are two types of these: the dispersive and the less known, but perhaps even more important entropic uncertainty relations. Below we state both the standard dispersive uncertainty relation and the version of the entropic uncertainty relation that concerns observables in finite dimensional Hilbert spaces.

OBSERVABLES AND STATES

Proposition 2.6 (Heisenberg Theorem) *Let X, Y and Z be selfadjoint operators on a Hilbert space $\mathcal{H}$ with domains $D(X), D(Y), D(Z)$, and let φ be a state given by the density matrix w. Assume furthermore that the following conditions hold:*

(i) *there exists a $D = D(X) \cap D(Y) \cap D(Z)$ dense in $\mathcal{H}$ linear subspace such that*

(ii)
$$(XY - YX)\eta = iZ\eta \qquad \eta \in D$$

(iii) *if one of X, Y and Z is not bounded, then w is a finite range operator $w = \sum_i r_i P_{\xi_i}$ such that $\xi_i \in D$; if X, Y and Z are bounded, then w is an arbitrary density matrix.*

If $\sigma_\varphi(X)$ and $\sigma_\varphi(Y)$ are the dispersions of φ on X, Y, and $\langle Z \rangle_\varphi$ is the expectation value of Z in φ (see the formulas (2.23)-(2.22)), then the following inequality holds

$$\sigma_\varphi(X)\sigma_\varphi(Y) \geq \frac{1}{4} | \langle Z \rangle_\varphi |^2 \qquad (2.26)$$

(Note that the conditions (i)-(iii) in the proposition are necessary because if they are not satisfied, then the quantities in (2.26) might not be well defined.)

It is very easy to prove the above uncertainty relation. Much more difficult is to demonstrate entropic uncertainty relations. Here we restrict ourselves to entropic uncertainty relations in finite dimensional Hilbert spaces; however, we give first the relevant definition of the entropic uncertainty in the case of not necessarily finite dimensional Hilbert spaces.

Definition 2.7 Let $\{d_i\}$ be a partition of real numbers into disjoint Borel measurable sets, and let X be a selfadjoint operator on the Hilbert space $\mathcal{H}$. The *entropic uncertainty* of the state ϕ on X with respect to the partition $\{d_i\}$ is given by

$$S(\phi, X, \{d_i\}) \equiv -\sum_i^\infty \phi(\mathrm{P}^X(d_i)) \log \phi(\mathrm{P}^X(d_i)) \qquad (2.27)$$

We are now in the position to state the following entropic uncertainty relation:

Proposition 2.7 *Let A, B be two selfadjoint operators on the finite dimensional Hilbert space $\mathcal{H}_n$ ($n \geq 2$) with non-degenerate spectrum and with eigenvectors ξ_i, η_j ($i, j = 1, \ldots n$). Let $\{d_i\}$ be a partition of the real numbers into disjoint Borel sets such that $\{d_i\}$ isolates the eigenvalues: each*

d_i contains one single eigenvalue. Then the following entropic ucertainty relation holds in any state ϕ

$$S(\phi, A, \{d_i\}) + S(\phi, B, \{d_i\}) \geq -2\log(\sup_{i,j}\{|\langle \xi_i, \eta_j \rangle|\}) \qquad (2.28)$$

Proof: It is enough to prove the inequality (2.28) for an arbitrary pure (vector) state given by the vector $\Phi \in \mathcal{H}_n$, since the left hand side of (2.28) is concave in ϕ. The proof is based on the following Riesz-Thorun convexity theorem:

Proposition 2.8 (Riesz-Thorun Theorem) *Let $T: \mathbb{C}^n \to \mathbb{C}^n$ be a linear map, and for $0 < p, q \leq \infty$ let*

$$d(p,q) \equiv \sup_{v \in C^n}\{\frac{\|Tv\|_p}{\|v\|_q}\}$$

where the L^p-norm of an element $v \in \mathbb{C}^n$ is defined by

$$\|v\|_p \equiv (\sum_i^n |v_i|^p)^{\frac{1}{p}}$$

($0 < p < \infty$), and

$$\|v\|_\infty \equiv \sup_i\{|v_i|\}$$

Then the two-place function

$$(x,y) \to \log(d(x^{-1}, y^{-1}))$$

is convex on the unit square $[0,1] \times [0,1]$ (with the convention $0^{-1} = \infty$).

Consider the matrix $T_{i,j} \equiv (\langle \xi_i, \eta_j \rangle)_{i,j}$. This T is a $\mathbb{C}^n \to \mathbb{C}^n$ linear operator, and below we apply the Riesz-Thorun theorem to T. $T_{i,j}$ is the matrix of the unitary transformation sending the orthonormal basis ξ_i into the basis η_i, therefore it is an isometry:

$$\|Tv\|_2 = \|v\|_2 \qquad (2.29)$$

Furthermore

$$\|Tv\|_\infty = \qquad (2.30)$$

$$= \sup_i |\sum_j \langle \xi_i, \eta_j \rangle v_j| \leq \sup_{i,j}\{|\langle \xi_i, \eta_j \rangle|\} \sum_j |v_j| \qquad (2.31)$$

$$= c\|v\|_1 \qquad (2.32)$$

where the notation
$$c \equiv \sup_{i,j}\{|\langle \xi_i, \eta_j \rangle|\}$$
was introduced. (2.29) and (2.30) can be expressed by $d(1/2,1/2) = 0$ and $d(0,1) = \log c$. We now apply the convexity property of $\log d$ to estimate $d(x, 1-x)$ $(0 < x < 1)$. Since

$$(1-x, x) = (2x-1)(0,1) + (2-2x)(1/2,1/2)$$

the convexity of $\log d$ implies that

$$\log d(1-x, x) \leq$$
$$(2x-1)\log d(0,1) + (2-2x)\log d(1/2,1/2)$$
$$= (2x-1)\log c$$

It follows that

$$\| Tv \|_{1/(1-x)} \leq d(1-x, x) \| v \|_{1/x} \quad (2.33)$$
$$\leq e^{(2x-1)\log c} \| v \|_{1/x} = c^{2x-1} \| v \|_{1/x} \quad (2.34)$$

Performing the parameter transformations

$$1/(1-x) \to p$$
$$(1/x) \to q$$
$$(2x-1) \to (1-2/p)$$

with $(1/p + 1/q = 1)$ (2.33) can be re-written as

$$\| Tv \|_p \leq c^{(1-2/p)} \| v \|_q \quad (2.35)$$

which implies

$$\log \| Tv \|_p \leq (1-2/p)\log c + \log \| v \|_q \quad (2.36)$$

Take now
$$v = (\langle \xi_1, \Phi \rangle, \langle \xi_2, \Phi \rangle, \ldots, \langle \xi_n, \Phi \rangle)$$
Then
$$Tv = (\langle \eta_1, \Phi \rangle, \langle \eta_2, \Phi \rangle, \ldots, \langle \eta_n, \Phi \rangle)$$
and the entropies can be written as

$$S(\Phi, A, \{d_i\}) = -\sum_i | (Tv)_i |^2 \log | (Tv)_i |^2$$

and
$$S(\Phi, B, \{d_i\}) = -\sum_i |v_i|^2 \log |v_i|^2$$

It also holds that
$$\frac{d \log \| Tv \|_p}{dp}\Big|_{p=2} = -\frac{1}{4} S(\Phi, A, \{d_i\})$$

and
$$\frac{d \log \| v \|_q}{dp}\Big|_{p=2} = \frac{1}{4} S(\Phi, B, \{d_i\})$$

indeed, one has
$$\frac{d \log \| Tv \|_p}{dp} = \frac{d}{dp}\left(\frac{1}{p} \log \sum_i |(Tv)_i|^p\right)$$
$$= -\frac{1}{p^2} \log \sum_i |(Tv)_i|^p + \frac{1}{p}\frac{d}{dp} \log \sum_i |(Tv)_i|^p$$
$$= -\frac{1}{p^2} \log \sum_i |(Tv)_i|^p$$
$$+ \frac{1}{p}\left(\sum_i |(Tv)_i|^p\right)^{-1} \sum_i |(Tv)_i|^p \log |(Tv)_i|$$

and since $\sum_i |(Tv)_i|^2 = 1$ it follows that
$$\frac{d \log \| Tv \|_p}{dp}\Big|_{p=2} = -\frac{1}{4} \log \sum_i |(Tv)_i|^2$$
$$+ \frac{1}{2}\left(\sum_i |(Tv)_i|^2\right)^{-1} \sum_i |(Tv)_i|^2 \log |(Tv)_i|$$
$$= \frac{1}{2} \sum_i |(Tv)_i|^2 \log |(Tv)_i|$$
$$= \frac{1}{4} \sum_i |(Tv)_i|^2 \log |(Tv)_i|^2 =$$
$$-\frac{1}{4} S(\Phi, A, \{d_i\})$$

and very similarly for $\frac{d \log \|v\|_q}{dp}\Big|_{p=2}$. Thus, one can estimate
$$\frac{d \log \| Tv \|_p}{dp}\Big|_{p=2} = \lim_{p \searrow 2} \frac{\log \| Tv \|_p}{p - 2}$$

from above by dividing the right hand side of (2.36) by $(p-2)$ and taking the limit $p \searrow 2$, and one obtains

$$-\frac{1}{4}S(\Phi, A, \{d_i\}) \leq \frac{1}{2}\log c + \frac{1}{4}S(\Phi, B, \{d_i\})$$

which is the inequality to be proved. □

Applying the above entropic uncertainty relation to complementary observables, the right hand side of (2.28) becomes $\log n$, hence we obtain as a corollary of Proposition 2.7

Proposition 2.9 *If A and B are complementary observables on the n-dimensional Hilbert space in the sense of Definition 2.7, then we have*

$$S(\phi, A, \{d_i\}) + S(\phi, B, \{d_i\}) \geq \log n \qquad (2.37)$$

for any state ϕ.

One of the important features of the entropic uncertainty relation described by Proposition 2.7 that distinguishes it from the dispersive uncertainty relation is that the lower bound on the sum of the entropies does not depend on the state.

Proposition 2.9 is the strongest possible entropic uncertainty relation in finite dimension, since one can always attain $\log n$ by chosing as state one of the eigenstates of A or B.

It is apparent that in the key concepts of the Hilbert space formalism (such as the definition of spectral measure, spectral theorem, Gleason's theorem) only the projections play a role, the elements of the Hilbert space themselves are secondary. This already indicates that it is the structure of the set of projections which is essential in quantum mechanics. Investigating the properties of the set $\mathcal{P}(\mathcal{H})$ of all projections will be the topic of the next chapter.

2.3. Bibliographic Notes

There are many textbooks on quantum mechanics available but only a few aim at a mathematically rigorous presentation. Von Neumann's classic work [168] is still strongly recommended, a more recent, mathematically minded treatment (using modern terminology and notation) is Prugovecki's monograph [115]. Gleason's theorem was proved in [52], and it is valid in the more general context of von Neumann algebras, see the review [88]. The Heisenberg theorem is one of the oldest results in quantum mechanics, for its history and interpretation see [159]. The Definition 2.7 of complementarity for finite dimensional observables is due to Schwinger [141]. The entropic uncertainty relations are far more sophisticated than the dispersive ones.

Proposition 2.7 was conjectured by Kraus [87], and was proved later by Maassen and Uffink in [96], where entropic uncertainty relations for possibly unbounded operators are also presented. The crucial Lemma in the proof of Proposition 2.7, the Riesz-Thorun interprolation theorem, is due to Marcel Riesz [137] (also see [63]). A slightly more general formulation of the discrete entropic uncertainty relation in terms of von Neumann algebras can be found in [95]. For a comprehensive review of entropic type uncertainty relations in quantum theory see [159] and [110].

CHAPTER 3

Lattice theoretic notions

In this chapter the basic notions of lattice theory are collected. Besides the definition of a lattice the important definitions are the *distributivity*, the *modularity* and the *orthomodularity* of a lattice (Definitions 3.5, 3.6 and 3.10). The proposition stating that a lattice on which a finite dimension function exists is necessarily modular (Proposition 3.3) will become important in the Chapter 6. The one-to-one correspondence between prime filters in a lattice and lattice homomorphisms on the lattice into a Boolean lattice (Proposition 3.11) will be used later to show that there exist no lattice homomorphisms from a quantum logic into a Boolean lattice. The notion of a partial algebra, partial Boolean algebra and partial algebra homomorphism will come up in Chapter 9 naturally in connection with a certain concept of hidden variable theory of quantum mechanics.

3.1. Basic notions in lattice theory

The pair $(\mathcal{L}, \leq)$ is a partially ordered set if a reflexive, transitive and antisymmetric relation (partial ordering) $\leq$ is defined between the elements of $\mathcal{L}$. If $A \leq B$ then we say "A is smaller than B", or "B is greater than A". This latter fact is also written as $B \geq A$. If $A \leq B$ but $A \neq B$ then we write $A < B$ and say "A is strictly smaller than B". The element $A \in \mathcal{L}$ is called an upper (lower) bound (with respect to the ordering $\leq$) of the set $\mathcal{S}$ of elements in $\mathcal{L}$ if $B \leq A$ (resp. $B \geq A$) for all $B \in \mathcal{S}$. The element A is the *least* upper bound (resp. *greatest* lower bound) of $\mathcal{S}$ if $A \leq A'$ (resp. $A \geq A'$) for any A' which is an upper (resp. lower) bound of $\mathcal{S}$. Since $\leq$ is an antisymmetric relation, any $\mathcal{S}$ has at most one least upper (resp. greatest lower) bound.

Definition 3.1 The partially ordered set $(\mathcal{L}, \leq)$ is called a *lattice* if for any two elements A, B there exists the least upper bound denoted by $A \vee B$ and the greatest lower bound denoted by $A \wedge B$ of A and B. The lattice is said to have zero and unit elements if there are elements $0, I$ in $\mathcal{L}$ such that $0 \leq A$ and $A \leq I$ for every $A \in \mathcal{L}$.

In what follows every lattice will be assumed to have a zero and unit elements, such lattices are called *bounded*.

Proposition 3.1 *In a lattice $\mathcal{L}$ the following equalities hold*

$$A \wedge A = A \qquad A \vee A = A \qquad (3.1)$$
$$A \wedge B = B \wedge A \qquad A \vee B = B \vee A \qquad (3.2)$$
$$A \wedge (B \wedge C) = (A \wedge B) \wedge C \qquad A \vee (B \vee C) = (A \vee B) \vee C \qquad (3.3)$$
$$A \wedge (A \vee C) = A \qquad A \vee (A \wedge B) = A \qquad (3.4)$$

furthermore, the above properties (i)-(iv) determine the lattice completely.
(The properties above are called *idempotency* (3.1), *commutativity* (3.2), *associativity* (3.3) and *absorption* (3.4).)

Proof: It is easy to check (i)-(iv) just by using the definition of lattice. If the operations $\wedge, \vee$ having the properties (i)-(iv) are defined on $\mathcal{L}$, then let a $\leq$ relation be defined by

$$A \leq B \text{ if and only if } A = A \wedge B$$

This relation is easily seen to be a partial ordering on $\mathcal{L}$. Since

$$A \wedge (A \wedge B) = ((A \wedge A) \wedge B) = (A \wedge B)$$

the element $A \wedge B$ is a lower bound (with respect to the ordering just defined) of A and, because of the commutativity of $\wedge$, also of B. What is more, $A \wedge B$ is the greatest lower bound of A and B, since, if $A \geq C$ and $B \geq C$, then

$$(A \wedge B) \wedge C = A \wedge (B \wedge C) = A \wedge C = C$$

i.e. $A \wedge B \geq C$. One can similarly reason by replacing $\wedge$ by $\vee$. $\square$

Definition 3.2 The lattice $\mathcal{L}$ is called a *complete* lattice if any subset of $\mathcal{L}$ has a greatest lower and a least upper bound. The lattice $\mathcal{L}$ is called σ-*lattice* if any countable subset of $\mathcal{L}$ has a greatest lower and least upper bound.

Example: Let X be an arbitrary set. The power set $Exp(X)$ (the set of all subsets of X) is a partially ordered set with respect to the set'theoretical inclusion and $Exp(X)$ also is a complete lattice: the least upper bound of a $\mathcal{S} \subset Exp(X)$ is given by the union of the members of $\mathcal{S}$, the greatest lower bound of $\mathcal{S}$ is the intersection of the members of $\mathcal{S}$.

Definition 3.3 The element $A \in \mathcal{L}$ is an *atom* in $\mathcal{L}$ if $B \leq A$ implies $B = A$ or $B = 0$. The lattice $\mathcal{L}$ is called an *atomic* lattice if for any $B \in \mathcal{L}$ there exists an atom A such that $A \leq B$. The lattice is called *completely* *atomistic* if any element is equal to the least upper bound of all the atoms it majorizes, i.e. if for any $0 \neq A \in \mathcal{L}$ it holds that

$$B = \vee_i A_i, \qquad A_i \leq B$$

where A_i is atom.

Definition 3.4 The series A_i $(i = 1, \ldots n)$ of lattice elements is called a *chain* if $A_1 \neq 0, A_i \neq A_j$ $(i \neq j)$ and

$$A_1 \leq A_2 \leq \ldots A_n$$

The number n is the *length* of the chain. The *rank* of a lattice is the supremum of the lengths of all possible chains.

Definition 3.5 The lattice $\mathcal{L}$ is called *distributive* if the following equality holds

$$A \vee (B \wedge C) = (A \vee B) \wedge (A \vee C) \quad for \ any \quad A, B, C \in \mathcal{L} \qquad (3.5)$$

Definition 3.6 The lattice $\mathcal{L}$ is called *modular* if the following equality holds

$$if \quad A \leq B \quad then \quad A \vee (B \wedge C) = (A \vee B) \wedge (A \vee C) \qquad (3.6)$$

If $A \leq B$ then $A \vee B = B$, and so the modularity equality (3.6) is equivalent to

$$if \quad A \leq B \quad then \quad A \vee (B \wedge C) = B \wedge (A \vee C) \qquad (3.7)$$

It is clear that every distributive lattice is modular, since modularity is a weakening of the distributivity property. The lattice $Exp(X)$ described above is distributive but there exist non-distributive modular lattices, as we shall see later.

Our next aim is to prove Proposition 3.3, which gives a sufficient condition for a lattice to be modular in terms of the dimension function on the lattice. As a preparation for this proposition we first define polinoms in a lattice, properties of which lead to a simplification of proving modularity in any lattice.

The polinoms are defined by induction as follows. Let $\mathcal{L}$ be a lattice and let x_i $(i = 1, 2, \ldots n)$ denote a finite set of variables that refer to elements in $\mathcal{L}$. The set of n-place polinoms contains, by definition, the terms

$$f(x_1, x_2 \ldots x_n) = x_i \quad \text{for some } i$$

and if

$$f_1(x_1, x_2 \ldots x_n) \text{ and } f_2(x_1, x_2 \ldots x_n)$$

are n-place lattice polinoms, then

$$f_1(x_1, x_2 \ldots x_n) \quad \wedge \quad f_2(x_1, x_2 \ldots x_n)$$
$$f_1(x_1, x_2 \ldots x_n) \quad \vee \quad f_2(x_1, x_2 \ldots x_n)$$

are defined also to be n-place lattice polinoms.

Proposition 3.2 *If f is a lattice polinom, then it holds that if*

$$x_i \leq y_i \quad (i = 1, 2, \ldots n)$$

then

$$f(x_1, x_2 \ldots x_n) \leq f(y_1, y_2 \ldots y_n)$$

(This property is expressed by saying that every lattice polinom is an isoton function *of the variables it contains.)*

Proof: It suffices to show that if $x \leq x'$, then

$$x \wedge y \leq x' \wedge y$$

and

$$x \vee y \leq x' \vee y$$

This is true, since one can write

$$x \wedge y = (x \wedge x') \wedge y = x \wedge (x' \wedge y) \leq x' \wedge y$$

and similarly for $\vee$. □

As a consequence of the above proposition the "one-sided distributive laws" hold, i.e. for any three elements A, B, C in any lattice we have that

$$\begin{aligned} A \vee (B \wedge C) &\leq (A \vee B) \wedge (A \vee C) \\ A \wedge (B \vee C) &\geq (A \wedge B) \vee (A \wedge C) \end{aligned} \quad (3.8)$$

Consequently, also the "one-sided modularity" always holds in any lattice:

$$if \quad A \leq B \quad then \quad A \vee (B \wedge C) \leq (A \vee C) \wedge B \quad (3.9)$$

Thus to prove the modularity of a lattice it suffices to show that

$$if \quad A \leq B \quad then \quad A \vee (B \wedge C) \geq (A \vee C) \wedge B \quad (3.10)$$

Definition 3.7 The map d defined on a lattice $\mathcal{L}$ and taking on non-negative (possibly infinite) values is called *dimension function* if it has the following properties:

(i) $d(A) < d(B)$ if $A < B$
(ii) $d(A) + d(B) = d(A \vee B) + d(A \wedge B)$

Proposition 3.3 *If there exists a dimension function d on the lattice $\mathcal{L}$ which takes on finite values, then the lattice is modular.*

Proof: It suffices to show that (3.10) holds, i.e. we must show that if $A \leq B$, then

$$A \vee (B \wedge C) < (A \vee C) \wedge B \qquad (3.11)$$

cannot hold. If (3.11) did hold, then

$$d(A \vee (B \wedge C)) < d((A \vee C) \wedge B) \qquad (3.12)$$

would follow; however, using the properties (i)-(ii) of d it is easy to see that (3.12) cannot be the case:

$$d(A \vee (B \wedge C)) - d((A \vee C) \wedge B) =$$
$$d(A) + d(B \wedge C) - d(A \wedge (B \wedge C))$$
$$-d(B) - d(A \vee C) + d((A \vee C) \vee B)$$
$$= d(A) + d(B \wedge C) - d(A \wedge C)) \quad - \quad d(B) - d(A \vee C) + d(B \vee C)$$
$$= d(A) - d(A \wedge C) - d(A \vee C)) \quad - \quad \bigl(d(B) - d(B \wedge C) - d(B \vee C)\bigr)$$
$$= d(C) - d(C) = 0$$

□

Definition 3.8 Let $\mathcal{L}$ be a lattice. The map

$$A \mapsto A^\perp$$

is called *orthocomplementation* and $A^\perp$ is called the *orthocomplement* of A if it has these properties:
(i) $(A^\perp)^\perp = A$
(ii) If $A \leq B$ then $B^\perp \leq A^\perp$
(iii) $A \wedge A^\perp = 0$
(iv) $A \vee A^\perp = I$

If an orthocomplementation is defined on a lattice $\mathcal{L}$, then the lattice $\mathcal{L}$ is called an *orthocomplemented* lattice. If A and B are elements in an orthocomplemented lattice, then they are called *orthogonal* if $A \leq B^\perp$.

Definition 3.9 A *Boolean algebra* is an orthocomplemented, distributive lattice. A Boolean algebra which is also a σ-lattice is called *Boolean σ-algebra*.

Example: In the lattice $Exp(X)$ the
$$A^\perp \equiv (X \setminus A)$$
is an orthocomplementation and $Exp(X)$ is a Boolean σ-algebra.

Example: The map $\mu^*: Exp(X) \to \mathbb{R}_+$ is called an *outer measure* if it has the properties:
(i) $\mu^*(0) = 0$
(ii) $\mu^*(A_1) \leq \mu^*(A_2)$ if $A_1 \subseteq A_2$
(iii) $\mu^*(\vee_n A_n) \leq \sum_n \mu^*(A_n)$ for arbitrary, countable A_n.
$B \in Exp(X)$ is called a *measurable* set with respect to μ^*, if
$$\mu^*(A) = \mu^*(A \wedge B) + \mu^*(A \setminus B)$$
for all $A \in Exp(X)$. The μ^*-measurable sets form a Boolean sub-σ-algebra in $Exp(X)$.

Example: Let V be a *finite dimensional* linear space and let $\mathcal{L}(V)$ be the set of all linear subspaces of V. Then $\mathcal{L}(V)$ is a *modular* lattice with the set theoretical inclusion as partial ordering and the set theoretical intersection as $\wedge$ and $A \vee B$ as the sum of A and B:
$$A \vee B = \{\eta + \xi \mid \eta \in A, \xi \in B\} \tag{3.13}$$
(the direct proof of modularity is the same as the proof of Proposition 4.3). $\mathcal{L}(V)$ is called the *projective geometry* determined by V. The lattice $\mathcal{L}(V)$ is not distributive if V is at least two dimensional. (If V happens to be a Hilbert space, then this follows from Proposition 4.16.) The rank of $\mathcal{L}(V)$ is equal to the dimension of V, and $\mathcal{L}(V)$ is an *atomic* lattice: the atoms are the one dimensional subspaces, which are also called the *points* of the projective geometry. The map d defined by
$$d(V_0) \equiv \dim(V_0)$$
($V_0 \subseteq V$ linear subspace) is a dimension function on $\mathcal{L}(V)$ in the sense of the Definition 3.7 and it has finite values; thus the modularity of $\mathcal{L}(V)$ follows from Proposition 3.3 too.

Definition 3.10 An orthocomplemented lattice is called an *orthomodular* lattice if it holds that
$$if \quad A \leq B \quad and \quad A^\perp \leq C \quad then \quad A \vee (B \wedge C) = (A \vee B) \wedge (A \vee C)$$

The orthomodularity property is a weakening of the modularity property, thus it is a further weakening of the distributivity property; one has the implication distributivity ⇒ modularity ⇒ orthomodularity

In Chapter 4 we shall see that there are indeed modular but not distributive lattices, and that there exist lattices that are orthomodular but not modular.

Proposition 3.4 *The following (so-called "De Morgan rules") hold in an orthocomplemented lattice:*

$$(\vee_n A_n)^\perp = \wedge_n A_n^\perp \quad (3.14)$$
$$(\wedge_n A_n)^\perp = \vee_n A_n^\perp \quad (3.15)$$

Proof: Putting $A_n^\perp$ in place of A_n in (3.15) and taking the complement of both sides one obtains (3.14), thus it siffices to show (3.14) only. For all j it holds that

$$A_j \leq \vee_n A_n$$

thus

$$(\vee_n A_n)^\perp \leq A_j^\perp$$

by property (ii) of the orthocomplementation, and so

$$(\vee_n A_n)^\perp \leq \wedge_j A_j^\perp$$

To see the converse inequality consider the inequality

$$\wedge_n A_n^\perp \leq A_j^\perp$$

This implies

$$A_j \leq (\wedge_n A_n^\perp)^\perp$$

It follows that

$$\vee_j A_j \leq (\wedge_j A_j^\perp)^\perp$$

and, taking the complement again, one obtains

$$\wedge_j A_j^\perp \leq (\vee_j A_j)^\perp$$

□

Proposition 3.5 *In an orthocomplemented lattice $\mathcal{L}$ the conditions below are equivalent.*
(0) *Orthomodularity:*

$$\text{if } A \leq B \text{ and } A^\perp \leq C \text{ then } A \vee (B \wedge C) = (A \vee B) \wedge (A \vee C)$$

(i) *Short form of orthomodularity:*

$$\text{if } A \leq B \text{ then } B = A \vee (A^\perp \wedge B)$$

(i') *Dual of the short form of orthomodularity:*

$$\text{if } B \leq A \text{ then } B = A \wedge (A^\perp \vee B)$$

(ii) *Quasimodularity (also called weak modularity):*

$$\text{if } A \leq B \text{ and } A^\perp \leq C \text{ then } A \vee (B \wedge C) = B$$

(ii') *Dual form of quasimodularity:*

$$\text{if } B \leq A \text{ and } C \leq A^\perp \text{ then } A \wedge (B \vee C) = B$$

Proof: The equivalence of (i) and (i') and of (ii) and (ii') is an easy consequence of the properties of orthocomplementation and of the De Morgan rules. Putting $A^\perp$ in place of C in (ii') one obtains (i'). To prove (i') $\Rightarrow$ (ii') note that if $B \leq A$ and $C \leq A^\perp$, then

$$B \vee C \leq B \vee A^\perp \text{ and } A \wedge (B \vee C) \leq A \wedge (B \vee A^\perp)$$

Assuming (i') it follows that

$$A \wedge (B \vee C) \leq B$$

and because of $B \leq A$ it holds that

$$B \leq A \wedge (B \vee C)$$

It remains to be seen that (0) and (i) are equivalent. (i) follows from (0) by putting $C = A^\perp$ into (0) and noting that $A \vee B = B$ due to $A \leq B$:

$$A \vee (B \wedge A^\perp) = (A \vee B) \wedge (A \vee A^\perp) = B \wedge I = B$$

Since the one-sided distributivity

$$A \vee (B \wedge C) \leq (A \vee B) \wedge (A \vee C)$$

holds (see eq. (3.8)), to prove (i) $\Rightarrow$ (0) it is enough to see that

$$\text{if } A \leq B \text{ and } B = A \vee (A^\perp \wedge B)$$

then

$$A \vee (B \wedge C) \geq [A \vee (A^\perp \wedge B)] \wedge (A \vee C) \qquad (3.16)$$

holds if $A^\perp \leq C$. Since $A^\perp \leq C$ it follows that

$$A^\perp \vee A \leq C \vee A$$

which implies

$$A \vee C = I$$

thus (3.16) holds if

$$A \vee (B \wedge C) \geq A \vee (A^\perp \wedge B) \tag{3.17}$$

holds. But (3.17) is true, since $B \wedge C \geq A^\perp \wedge B$ due to $C \geq A^\perp$. □

In view of the above proposition, saying that an orthocomplemented lattice is orthomodular is the same as saying that it is quasimodular, which is the same as saying that it is weakly modular. In what follows the term "orthomodularity" will be used consistently.

Definition 3.11 The subset $\mathcal{S}$ in a lattice $\mathcal{L}$ is called *upwardly closed*, or *upper* (respectively *downward closed* or *lower*) if for all $A, B \in \mathcal{L}$ the fact $B \geq A \in \mathcal{L}$ entails $B \in \mathcal{L}$ ($B \leq A \in \mathcal{L}$ entails $B \in \mathcal{L}$).

For an arbitrary $\mathcal{S}$ let $\mathcal{S} \uparrow$ ($\mathcal{S} \downarrow$) denote the upper (lower) set generated by $\mathcal{S}$:

$$\mathcal{S} \uparrow \equiv \{X \in \mathcal{L} \mid \text{there is } A \in \mathcal{S} \text{ with } X \geq A\}$$
$$\mathcal{S} \downarrow \equiv \{X \in \mathcal{L} \mid \text{there is } A \in \mathcal{S} \text{ with } X \leq A\}$$

Definition 3.12 The subset $\mathcal{F}$ in a lattice $\mathcal{L}$ is called *filter* if $\mathcal{F}$ is an upper set and it is closed with respect to finite meets. A filter $\mathcal{F}$ is said to be *proper* if $\mathcal{F}$ is a proper subset of $\mathcal{L}$.

Definition 3.13 The subset $\mathcal{J}$ in a lattice $\mathcal{L}$ is called *ideal* if $\mathcal{J}$ is a lower set and it is closed with respect to finite union. An ideal $\mathcal{J}$ is said to be *proper* if $\mathcal{J}$ is a proper subset of $\mathcal{L}$.

Clearly, a filter is proper if and only if it does not contain 0, and an ideal is proper if it does not contain the unit element I of the lattice. Since the intersection of an arbitrary set of filters is again a filter, every set $\mathcal{S}$ of lattice elements generates a smallest filter that contains $\mathcal{S}$. This smallest filter is denoted by $\mathcal{F}(\mathcal{S})$.

Definition 3.14 A proper filter $\mathcal{F}$ in the lattice $\mathcal{L}$ is called
(i) *principal filter* if $\mathcal{F} = \{A\} \uparrow$ for a some element $A \in \mathcal{L}$
(ii) *maximal* (or *ultra*) *filter* if there exist no filter $\mathcal{F}' \supset \mathcal{F}$
(iii) *prime filter* if for all $A, B \in \mathcal{L}$ such that $(A \vee B) \in \mathcal{F}$ we have either $A \in \mathcal{F}$ or $B \in \mathcal{F}$.

Definition 3.15 A proper ideal $\mathcal{J}$ is called
(i) *principal* ideal if $\mathcal{J} = \{A\} \downarrow$ for some $A \in \mathcal{L}$
(ii) *maximal* if there exists no ideal $\mathcal{J}' \supset \mathcal{J}$
(iii) *prime* ideal if for all $A, B \in \mathcal{L}$ such that $(A \wedge B) \in \mathcal{F}$ we have either $A \in \mathcal{F}$ or $B \in \mathcal{F}$.

The notions of filter and ideal are dual notions. It also holds that

Proposition 3.6 *$\mathcal{F}$ is a prime filter if and only if $\mathcal{L} \setminus \mathcal{F}$ is a prime ideal.*

Proof: Obvious. □

The next proposition gives a characterization of maximal filters.

Proposition 3.7 *A filter $\mathcal{F}$ is maximal if and only if for every $B \notin \mathcal{F}$ there exists an $A \in \mathcal{F}$ such that $A \wedge B = 0$.*

Proof: If $A \wedge B \neq 0$ for every $A \in \mathcal{F}$, then the filter defined by

$$\{A \wedge B \mid A \in \mathcal{F}\} \uparrow$$

is clearly a filter larger than $\mathcal{F}$, hence $\mathcal{F}$ is not a maximal filter. Conversely, if $\mathcal{F}'$ is a filter larger than $\mathcal{F}$, then for any $B \in \mathcal{F}' \setminus \mathcal{F}$ and any $A \in \mathcal{F}$ one has $A \wedge B \in \mathcal{F}'$, therefore $A \wedge B \neq 0$. □

Proposition 3.8 *If $\mathcal{L}$ is a distributive lattice, then every maximal filter is prime.*

Proof: Let $\mathcal{F}$ be a maximal filter in the distributive lattice. We show that if $\mathcal{L} \ni A, B \notin \mathcal{F}$ then $A \vee B \notin \mathcal{F}$. By Proposition 3.7 there exist elements $C, D \in \mathcal{F}$ such that
$$A \wedge C = B \wedge D = 0$$
Let
$$E = C \wedge D \in \mathcal{F}$$
Then $C \neq 0$ because $\mathcal{F}$ is proper, furthermore, by the distributivity of $\mathcal{L}$ we can write

$$E \wedge (A \vee B) = (E \wedge A) \vee (E \wedge B) =$$
$$((C \wedge D) \wedge A) \vee ((C \wedge D) \wedge B) = 0 \vee 0 = 0$$

It follows that $(A \vee B) \notin \mathcal{F}$, for if $(A \vee B) \in \mathcal{F}$ then $C \wedge (A \vee B) = 0$ also is in $\mathcal{F}$ because $\mathcal{F}$ is a filter. But 0 can not be in $\mathcal{F}$ because $\mathcal{F}$ is proper. □

Proposition 3.9 *Let $\mathcal{L}$ be a Boolean algebra and $\mathcal{F}$ be a proper filter in $\mathcal{L}$. Then the following conditions are equivalent:*

(i) $\mathcal{F}$ is maximal
(ii) $\mathcal{F}$ is prime
(iii) For every $A \in \mathcal{L}$ either $A \in \mathcal{F}$ or $A^\perp \in \mathcal{F}$

Proof: The implication (i)⇒(ii) is the content of Proposition 3.8. Suppose (ii) holds. Since $A \vee A^\perp = I \in \mathcal{F}$, either $A \in \mathcal{F}$ or $A^\perp \in \mathcal{F}$. Assume that (iii) holds and that $A \notin \mathcal{F}$. Then there exists a B (namely $B = A^\perp$) for which $A \wedge B = 0$, hence by Proposition 3.7 $\mathcal{F}$ is maximal. □

Definition 3.16 The map $h: \mathcal{L}_1 \to \mathcal{L}_2$ between two lattices $\mathcal{L}_1, \mathcal{L}_2$ is called a *lattice homomorphism* if

$$h(A \vee B) = h(A) \vee h(B) \tag{3.18}$$
$$h(A \wedge B) = h(A) \wedge h(B) \tag{3.19}$$

for all $A, B \in \mathcal{L}_1$. A lattice homomorphism is called *imbedding* if $A \neq B$ implies $h(A) \neq h(B)$.

Proposition 3.10 Let $h: \mathcal{L}_1 \to \mathcal{L}_2$ be a lattice homomorphism, and $\mathcal{F}$ be a filter in $\mathcal{L}_2$. Then
(i) $h^{-1}(\mathcal{F})$ is a filter in $\mathcal{L}_1$,
(ii) if $\mathcal{F}$ is a prime filter, then $h^{-1}(\mathcal{F})$ also is a prime filter.

Proof: Obvious. □

In the next proposition $\mathcal{L}_2$ denotes the two element lattice $\{0, I\}$.

Proposition 3.11 If $\mathcal{F}$ is a subset in a lattice $\mathcal{L}$, then the following are equivalent
(i) $\mathcal{F}$ is a prime filter
(ii) $\mathcal{F} = h^{-1}(I)$ for some lattice homomorphism $h: \mathcal{L} \to \mathcal{L}_2$.

Proof: The implication (ii)⇒(i) is a consequence of Proposition 3.10 because $\{I\}$ is a filter. To see (i)⇒(ii) assume that $\mathcal{F}$ is a prime filter, and define $h: \mathcal{L} \to \mathcal{L}_2$ by

$$h(A) = \begin{cases} I, & \text{if } A \in \mathcal{F} \\ 0, & \text{if } A \notin \mathcal{F} \end{cases}$$

Obviously $\mathcal{F} = h^{-1}(I)$, so one has to check the homomorphism properties of h only. This is routine: we must show that

$$h(A \vee B) = h(A) \vee h(B) \tag{3.20}$$
$$h(A \wedge B) = h(A) \wedge h(B) \tag{3.21}$$

If $(A \vee B) \in \mathcal{F}$, then the left hand side of eq. (3.20) is equal to I, but then either $A \in \mathcal{F}$ or $B \in \mathcal{F}$ by the prime ideal property of $\mathcal{F}$, so either $h(A) = I$, or $h(B) = I$, i.e. the right hand side of eq. (3.20) is equal to I. If $(A \vee B) \notin \mathcal{F}$, then the left hand side of eq. (3.20) is equal to 0, and in this case both $A \notin \mathcal{F}$ and $B \notin \mathcal{F}$ hold because $\mathcal{F}$ is upper closed and $A \leq (A \vee B)$, so $A \in \mathcal{F}$ or $B \in \mathcal{F}$ would imply $(A \vee B) \in \mathcal{F}$, contrary to the assumption. All this means that the right hand side of eq. (3.20) is equal to 0 in this case. If $(A \wedge B) \in \mathcal{F}$ then $h(A \wedge B) = I$ and if in this case we had $h(A) \wedge h(B) = 0$, then either $h(A)$ or $h(B) = 0$ would hold, which would imply either $A \notin \mathcal{F}$ or $B \notin \mathcal{F}$, but this cannot be since $(A \wedge B) \in \mathcal{F}$, $(A \wedge B) \leq A$, $(A \wedge B) \leq B$ and $\mathcal{F}$ is upper closed. If $(A \wedge B) \notin \mathcal{F}$ then $h(A \wedge B) = 0$, and if in this case we had $h(A) \wedge h(B) = I$, then $A, B \in \mathcal{F}$ would follow, but then $(A \wedge B) \in \mathcal{F}$, since $\mathcal{F}$ is closed with respect to the meet operation. □

Proposition 3.11 tells us that there is a one-to-one correspondence between prime filters (and prime ideals) in a lattice and the two-valued homomorphisms on the lattice. Since a two valued homomorphism h on a lattice $\mathcal{L}$ is just a *valuation map*, i.e. a map that assigns either the value "truth" I or the value "false" 0 to the elements in $\mathcal{L}$, the message of the Proposition 3.11 can also be worded by saying that there exists a valuation map on a lattice if and only if the lattice contains a prime filter.

We shall see in Section 4.1 that the lattice of projections on a Hilbert space of dimension 3 or greater does not have a prime filter, hence those lattices do not allow a valuation map (see Proposition 4.9).

Our next aim is to define the so-called partial Boolean algebras. One way to do this is to define them as idempotent elements in a partial algebra, so we first define the latter ones. The idea of the notion of partial algebra is to restrict the meaningfulness of the algebraic operations to a certain subset of the possible algebraic elements, where the subset is singled out by introducing a relation † "commeasurability" and the algebraic operations will be defined between those elements only that stand in this commeasurablity relation.

Definition 3.17 The pair $(\mathcal{A}, \dagger)$ is a partial algebra over the field K (which we always assume to be either $\mathbb{R}$ or the complex numbers) if $\dagger \subseteq \mathcal{A} \times \mathcal{A}$ is a binary relation on $\mathcal{A}$ (called "commeasurabilty") and for every pair $(A, B) \in \dagger$ a product operation

$$(A, B) \mapsto AB$$

a sum operation

$$(A, B) \mapsto (A + B)$$

and for every pair $(\lambda, A) \in K \times \mathcal{A}$ the multiplication by scalars

$$(\lambda, A) \mapsto \lambda A \in \mathcal{A}$$

are defined in such a way that the following conditions are satisfied

(i) † is reflexive and symmetric
(ii) There exists an element (unit) $I \in \mathcal{A}$ for which $A \dagger I$ for all $A \in \mathcal{A}$.
(iii) If $A_1, A_2, A_3 \in \mathcal{A}$ are pairwise commeasurable, then

$$(A_1 + A_2) \quad \dagger \quad A_3 \tag{3.22}$$
$$(A_1 A_2) \quad \dagger \quad A_3 \tag{3.23}$$
$$\lambda A_i \quad \dagger \quad A_3 \quad (i = 1, 2) \tag{3.24}$$

(iv) If $A_1, A_2, A_3 \in \mathcal{A}$ are pairwise commeasurable, then the polynomials in A_1, A_2, A_3 form a commutative algebra over K.

An example of a partial algebra, after which the above definition was in fact designed, is the set of all bounded operators $\mathcal{B}(\mathcal{H})$ on a Hilbert space with $A \dagger B$ being the commutation relation ($A \dagger B$ if and only if $AB = BA$), and the product and multiplication by scalars being the usual algebraic operations with respect to which $\mathcal{B}(\mathcal{H})$ is a (non-commutative) algebra (see Chapter 6).

Let $(\mathcal{A}, \dagger)$ be a partial algebra and let $\mathcal{B}$ be the subset in $\mathcal{A}$ of elements A such that $AA = A$. If for $A \dagger B$ we define

$$A \vee B \equiv (A + B) - AB$$
$$A \wedge B \equiv AB$$
$$A^\perp \equiv I - A$$
$$0 \equiv 0I$$

and consider the restriction of † to $\mathcal{B}$, then for the pair the $(\mathcal{B}, \dagger)$ following hold

(i) † is reflexive and symmetric
(ii) There exists an element (unit) $I \in \mathcal{B}$ for which $A \dagger I$ for all $A \in \mathcal{B}$.
(iii) If $A_1, A_2, A_3 \in \mathcal{B}$ are pairwise commeasurable, then

$$(A_1 \vee A_2) \quad \dagger \quad A_3 \tag{3.25}$$
$$A_1 A_2 \quad \dagger \quad A_3 \tag{3.26}$$
$$A_i^\perp \quad \dagger \quad A_j \quad (i \neq j = 1, 2) \tag{3.27}$$

(iv) If $A_1, A_2, A_3 \in \mathcal{B}$ are pairwise commeasurable, then the $(\vee, \wedge)$-polynomials in A_1, A_2, A_3 form a Boolean algebra.

Definition 3.18 The pair $(\mathcal{B}, \dagger)$ is called a *partial Boolean algebra* if the above conditions are satisfied.

An example of a partial Boolean algebra is the set of projection operators in $\mathcal{B}(\mathcal{H})$ with the commutativity relation as commeasurability.

Definition 3.19 The A, B elements in a partial Boolean algebra are said to be *orthogonal* $A \perp B$ if $A \dagger B$ and $A \wedge B = B \wedge A = 0$.

Definition 3.20 A map $h: \mathcal{A}_1 \to \mathcal{A}_2$ between two partial algebras $(\mathcal{A}_1, \dagger)$ and $(\mathcal{A}_2, \dagger)$ is called a *partial algebra homomorphism* if it has the following properties

(i) if $A \dagger B$ then $h(A) \dagger h(B)$
(ii) if $A \dagger B$ then

$$\begin{align} h(\lambda_1 A + \lambda_2 B) &= \lambda_1 h(A) + \lambda_2 h(B) \quad \lambda_1, \lambda_2 \in K & (3.28) \\ h(AB) &= h(A)h(B) & (3.29) \\ h(I) &= I & (3.30) \end{align}$$

If h is a partial algebra homomorphism from a partial algebra $(\mathcal{A}, \dagger)$ into a commutative algebra $\mathcal{C}$, then the restriction of h to the partial Boolean algebra $(\mathcal{B}, \dagger)$ determined by $(\mathcal{A}, \dagger)$ is a map from $\mathcal{B}$ into the Boolean algebra of idempotens of $\mathcal{C}$; furthermore we have: If A and B are orthogonal, then

$$h(A + B) = h(A) + h(B) = h(A) \vee h(B) \tag{3.31}$$

We shall see in the next chapter that there does not exist a partial algebra homomorphism from the partial Boolean algebra of quantum mechanics into a Boolean algebra (Proposition 4.10), and we shall see in Chapter 9 how this result relates to a certain concept of hidden variables.

For later references we give the following definitions.

Definition 3.21 Let $\mathcal{L}$ be an orthocomplemented lattice. The map

$$\phi: \mathcal{L} \to [0, 1]$$

is called *state* if it is additive on orthogonal elements, i.e. if

$$\phi(A \vee B) = \phi(A) + \phi(B) \qquad A, B \in \mathcal{L} \quad A \perp B$$

Definition 3.22 The state ϕ on a lattice $\mathcal{L}$ is called a *Jauch-Piron state* if the condition

$$\phi(A) = \phi(B) = 0$$

implies

$$\phi(A \vee B) = 0$$

Definition 3.23 The lattice $\mathcal{L}$ is a *Jauch-Piron lattice* if every state on $\mathcal{L}$ is a Jauch-Piron state.

3.2. Bibliographic notes

The standard reference for lattice theory is the classic work of Birkhoff [19], more recent monographs are [97] and [84]. Most textbooks on quantum logic also contain the definitions of the necessary lattice theoretic concepts. The notions of partial algebra, partial Boolean algebra and partial algebra homomorphism is taken from [85].

CHAPTER 4

Hilbert lattice

In this chapter the lattice properties of the set of closed linear subspaces of a (possibly infinite dimensional) complex Hilbert space are described. This lattice is an atomic, (completely) atomistic, complete, orthomodular lattice that has the (minimal) covering property. Proposition 4.3 says that the lattice of a *finite* dimensional Hilbert space is not only orthomodular but also modular, whereas the lattice of an infinite dimensional Hilbert space is *not* modular (Proposition 4.4). It will be seen that this important difference between the projection lattices of finite and infinite dimensional Hilbert spaces is related to the fact that a linear subspace of an infinite dimensional Hilbert space is not necessarily closed (see the Remark after Definition 4.3). It is proved that there exist no "evaluation map" on the Hilbert lattice if the dimension of the Hilbert space is greater than 2, not even if by an evaluation map one means the restriction of a partial algebra homomorphism from the partial algebra of bounded operators into a commutative algebra (Propositions 4.9 and 4.10). In the subsection 4.2 the relations are summed up that link the lattice operations to the algebraic operations defined between the projections. The two important statements in this subsection are that a sublattice is distributive if and only if it is formed by commuting projections (Proposition 4.16), and that a sublattice generated by mutually commuting projections is distributive (Proposition 4.17).

4.1. Hilbert space and the lattice of subspaces

Let $\mathcal{H}$ be a complex Hilbert space. $\mathcal{H}_0 \subset \mathcal{H}$ is called a *linear* subspace if $\mathcal{H}_0$ itself is a linear space i.e. if it is closed with respect to the sum in $\mathcal{H}$ and the multiplication with scalars. $\mathcal{H}_0$ is a *closed* linear subspace if it is a linear subspace *and* it is closed in the norm in $\mathcal{H}$, i.e. if it holds that if η_n is a Cauchy sequence in $\mathcal{H}_0$, then there is an $\zeta \in \mathcal{H}_0$ such that ζ is the limit of η_n. It is significant that in an infinite dimensional Hilbert space a linear subspace is not necessarily closed (see the Remark after Definition 4.3). A closed linear subspace $\mathcal{H}_0$ is itself a Hilbert space. $\mathcal{P}(\mathcal{H})$ denotes the set of all closed, linear subspaces of $\mathcal{H}$, and we use $\mathcal{H}_i$, and curly letters $\mathcal{G}, \mathcal{F}, \mathcal{E}$ to denote closed linear subspaces.

In what follows, we shall define certain operations between the elements of $\mathcal{P}(\mathcal{H})$, and we will see that $\mathcal{P}(\mathcal{H})$ becomes an orthomodular complete lattice having a number of additional properties.

Definition 4.1 (*Partial ordering in $\mathcal{P}(\mathcal{H})$*): Let $\leq$ be defined by: $\mathcal{H}_1 \leq \mathcal{H}_2$ if and only if $\mathcal{H}_1 \subseteq \mathcal{H}_2$; that is $\mathcal{H}_1 \leq \mathcal{H}_2$ if $\mathcal{H}_1$ as a set is contained in $\mathcal{H}_2$ as a set.

The relation $\leq$ is obviously reflexive, transitive and antisymmetric, i.e. it is a partial ordering in $\mathcal{P}(\mathcal{H})$.

Definition 4.2 (*Meet in $\mathcal{P}(\mathcal{H})$*): Let $\mathcal{G}_i$ be a family of closed linear subspaces. Their *meet* is defined to be the closed linear subspace

$$\wedge_i \mathcal{G}_i \equiv \cap_i \mathcal{G}_i$$

where $\cap$ is the set theoretical intersection.

Definition 4.3 (*Join in $\mathcal{P}(\mathcal{H})$*): Let $\mathcal{G}_i$ be a family of closed linear subspaces in $\mathcal{H}$. The *sum* of these subspaces is the linear subspace $\sum_i \mathcal{G}_i$ defined by

$$\sum_i \mathcal{G}_i = \{\eta \in \mathcal{H} \mid \eta = \sum_i^N \zeta_i, \ \zeta_i \in \mathcal{G}_i\} \qquad (4.1)$$

The *join*, which is denoted by $\vee_i \mathcal{G}_i$, is defined to be the *closed* linear subspace spanned by $\mathcal{G}_i$, i.e. $\vee_i \mathcal{G}_i$ is the norm-closure of $\sum_i \mathcal{G}_i$.

Remark: If $\mathcal{H}$ is not finite dimensional, then there are closed linear subspaces $\mathcal{E}, \mathcal{G}$ such that their sum defined by (4.1) is not closed. This can be seen in the following example: Let η_n, ξ_n ($n = 1, 2 \ldots$) be two infinite orthonormal sequences in $\mathcal{H}$ such that

$$\langle \eta_n, \xi_m \rangle = 0$$

holds for all n, m; furthermore, let $0 < a_n, b_n$ be two infinite series of positive real numbers such that the following two conditions hold

1. $\sum b_n^2 < \infty$
2. the set of vectors θ_n defined by

$$\theta_n \equiv a_n \eta_n + b_n \xi_n$$

(which is an orthogonal set of vectors) also is orthonormal, i.e.

$$1 = \| \theta_n \|^2 = a_n^2 + b_n^2$$

(Note that there are such numbers: $a_n = \cos(\frac{1}{n}), b_n = \sin(\frac{1}{n})$ would do for instance.) We claim that the two closed subspaces

$$\mathcal{E} = [\eta_n, n = 1, 2 \ldots]$$

and

$$\mathcal{G} = [\theta_n, n = 1, 2, \ldots]$$

are such that $\mathcal{E} \vee \mathcal{G}$ is not equal to the sum of $\mathcal{E}$ and $\mathcal{G}$. To see this, one must exhibit an element $\xi \in \mathcal{E} \vee \mathcal{G}$ such that $\xi \neq \eta^{\mathcal{E}} + \eta^{\mathcal{G}}$. Put $\xi = \sum b_n \xi_n$. The element ξ_n is in $\mathcal{E} + \mathcal{G}$ for every n since $b_n \neq 0$ and one can write

$$\xi_n = -\frac{a_n}{b_n}\eta_n + (\frac{1}{b_n}a_n\eta_n + \frac{1}{b_n}b_n\xi_n)$$

Since $\sum b_n < \infty$, it follows then that $\xi \in \mathcal{E} \vee \mathcal{G}$. If this ξ could be written as $\xi = \eta^{\mathcal{E}} + \eta^{\mathcal{G}}$, then it would follow that

$$\begin{aligned} b_n = \langle \xi, \xi_n \rangle &= \langle \eta^{\mathcal{E}} + \eta^{\mathcal{G}}, \xi_n \rangle \\ &= \langle \eta^{\mathcal{G}}, \xi_n \rangle = \langle \sum_j \langle \eta^{\mathcal{G}}, \theta_j \rangle \theta_j, \xi_n \rangle \\ &= \langle \eta^{\mathcal{G}}, \theta_n, \rangle b_n \end{aligned}$$

Since $b_n \neq 0$, it should follow that $\langle \eta^{\mathcal{G}}, \theta_n \rangle = 1$ for every n; however, this cannot be the case, since $\langle \eta^{\mathcal{G}}, \theta_n \rangle$ are the Fourier coefficients of the vector $\eta^{\mathcal{G}}$ with respect to θ_n.

Proposition 4.1 *We have the following*

(i) $\wedge_i \mathcal{G}_i$ *is the greatest lower bound of the subspaces $\mathcal{G}_i$ with respect to the ordering $\leq$, i.e. $\wedge_i \mathcal{G}_i \leq \mathcal{G}_i$ for all i, and if $\mathcal{F} \leq \mathcal{G}_i$ for all i, then $\mathcal{F} \leq (\wedge_i \mathcal{G}_i)$*

(ii) $\vee_i \mathcal{G}_i$ *is the least upper bound of the subspaces $\mathcal{G}_i$, i.e. $\vee_i \mathcal{G}_i \geq \mathcal{G}_i$ for all i, and if $\mathcal{F} \geq \mathcal{G}_i$ for all i, then $(\vee_i \mathcal{G}_i) \leq \mathcal{F}$*

Proof: Obvious

Definition 4.4 (*Orthocomplementation in $\mathcal{P}(\mathcal{H})$*): Let $\mathcal{G}$ be a closed linear subspace and define $\mathcal{G}^\perp$ by

$$\mathcal{G}^\perp \equiv \{\eta \in \mathcal{H} \mid \langle \eta, \zeta \rangle = 0 \text{ for all } \zeta \in \mathcal{G}\} \qquad (4.2)$$

$\mathcal{G}^\perp$ is a closed linear subspace, it is called the *orthogonal complement* of $\mathcal{G}$.

Proposition 4.2 *The map $\mathcal{G} \mapsto \mathcal{G}^\perp$ defined by (4.2) has the following properties*

(i) $(\mathcal{G}^\perp)^\perp = \mathcal{G}$

(ii) if $\mathcal{G} \leq \mathcal{F}$ then $\mathcal{F}^\perp \leq \mathcal{G}^\perp$
(iii) $\mathcal{G} \wedge \mathcal{G}^\perp = 0$
(iv) $\mathcal{G} \vee \mathcal{G}^\perp = I$

Proof: (i) and (ii) are obvious. To prove (iii) and (iv) it is enough to see one of them, since the other follows by the de Morgan rule. $\mathcal{G} \vee \mathcal{G}^\perp = I$ is just a re-writing in lattice theoretic notation of one of the elementary facts in Hilbert space theory, namely the fact that given any closed linear subspace $\mathcal{G}$ and any $\xi \in \mathcal{H}$ there exists a unique $\eta \in \mathcal{G}$ such that $(\xi - \eta) \in \mathcal{G}^\perp$. □

Proposition 4.3 *If the Hilbert space $\mathcal{H}$ is finite dimensional, then $\mathcal{P}(\mathcal{H})$ is a modular lattice.*

Proof: Let $\mathcal{F} \leq \mathcal{G}$. One must show that for every $\mathcal{E} \in \mathcal{P}(\mathcal{H})$ it holds that

$$\mathcal{F} \vee (\mathcal{E} \wedge \mathcal{G}) = (\mathcal{F} \vee \mathcal{E}) \wedge \mathcal{G}$$

If the subspaces are all finite dimensional, then

$$\eta \in (\mathcal{F} \vee (\mathcal{E} \wedge \mathcal{G}))$$

if and only if

$$\eta = \eta^{\mathcal{F}} + \eta^{\mathcal{E} \wedge \mathcal{G}}$$

(The superscripts indicating the subspace the vectors belong to.) But

$$\eta^{\mathcal{E} \wedge \mathcal{G}} \in (\mathcal{E} \wedge \mathcal{G})$$

if and only if

$$\eta^{\mathcal{E} \wedge \mathcal{G}} \in \mathcal{E} \quad \text{and} \quad \eta^{\mathcal{E} \wedge \mathcal{G}} \in \mathcal{G}$$

It follows that

$$\eta = (\eta^{\mathcal{F}} + \eta^{\mathcal{E} \wedge \mathcal{G}}) \in (\mathcal{F} \vee \mathcal{E})$$

since η it is the sum of vectors in $\mathcal{F}$ and in $\mathcal{E}$. Furthermore, the vector η also is in $\mathcal{G}$, since $\eta^{\mathcal{F}} \in \mathcal{G}$ by the assumption $\mathcal{F} \leq \mathcal{G}$. Conversely, if

$$\eta \in ((\mathcal{F} \vee \mathcal{E}) \wedge \mathcal{G})$$

then $\eta \in \mathcal{G}$ and $\eta \in (\mathcal{F} \vee \mathcal{E})$, which holds if and only if $\eta = \eta^{\mathcal{F}} + \eta^{\mathcal{E}}$, which implies (since $\eta \in \mathcal{G}$ and $\mathcal{F} \leq \mathcal{G}$)

$$\eta^{\mathcal{E}} = (\eta - \eta^{\mathcal{F}}) \in \mathcal{G}$$

and so $\eta^{\mathcal{E}} \in (\mathcal{F} \wedge \mathcal{G})$, consequently η is the sum of two vectors, one lying in $\mathcal{F}$, one lying in $\mathcal{E} \wedge \mathcal{G}$, i.e. $\eta \in \mathcal{F} \vee (\mathcal{E} \wedge \mathcal{G})$. □

It was essential in the above proof that the subspaces $\mathcal{E}, \mathcal{F}, \mathcal{G}$ are all finite dimensional. If this condition is not met, then the unions featuring in the modularity equality may contain elements that do not belong to either member of the union and these elements cannot be obtained as sums of elements belonging to the members in the union, as it was seen in the Remark following the Definition 4.3. Thus the above proof does not go through in the case of an infinite dimensional $\mathcal{H}$. But even more is true:

Proposition 4.4 *If $\mathcal{H}$ is not finite dimensional, then $\mathcal{P}(\mathcal{H})$ is not modular.*

Proof: Let η_n $(n = 1, 2, \ldots)$ be an orthonormed basis in $\mathcal{H}$, and let us define the elements ξ_n by

$$\xi_n = \eta_{2n} + a^{-n}\eta_1 + a^{-2n}\eta_{2n+1} \quad (a > 1)$$

Consider the subspaces $\mathcal{E}, \mathcal{F}, \mathcal{G}$ defined as follows:
$\mathcal{F} \equiv$ generated by the elements ξ_n $(n = 1, 2, \ldots)$
$\mathcal{G} \equiv$ generated by the elements ξ_n $(n = 1, 2, \ldots)$ and η_1
$\mathcal{E} \equiv$ generated by the elements η_{2n} $(n = 1, 2, \ldots)$

We show that the assumption of validity for $\mathcal{F}, \mathcal{G}$ and $\mathcal{E}$ of the modularity equality (3.6) leads to contradiction. Obviously

$$\mathcal{F} \wedge \mathcal{E} = 0 \quad \text{and} \quad \mathcal{E} \wedge \mathcal{G} = 0$$

Furthermore it holds that $\mathcal{F} \leq \mathcal{G}$, since all the elements $(\sum_n \lambda_n \xi_n) \in \mathcal{F}$ are also in $\mathcal{G}$. What is more, $\mathcal{F}$ is strictly smaller than $\mathcal{G}$ because

$$\xi'_n = \eta_{2n} + a^{-n}\eta_1 + a^{-2n}\eta_{2n+1} - a^{-n}\eta_1 = (\eta_{2n} + a^{-2n}\eta_{2n+1}) \notin \mathcal{F}$$

for any n; however, $\xi'_n \in \mathcal{G}$ for any n. Since $\mathcal{F} \leq \mathcal{G}$, it holds that

$$(\mathcal{F} \vee \mathcal{E}) \leq (\mathcal{G} \vee \mathcal{E})$$

The subspace $\mathcal{G}$ is the closure of the linear subspace of elements

$$(\sum_{n=1}^{N} \lambda_n \xi_n + \lambda \eta_1)$$

and the subspace $\mathcal{E}$ is the closure of the linear subspace of the elements

$$(\sum_{n=1}^{N} \lambda_n \eta_{2n})$$

But the elements
$$(\sum_{n=1}^{N} \lambda_n \xi_n + \lambda \eta_1)$$
are in $(\mathcal{F} \vee \mathcal{E})$, since
$$(\sum_{n=1}^{N} \lambda_n \xi_n) \in \mathcal{F}$$
and the vector η_1 also is in $(\mathcal{F} \vee \mathcal{E})$ because $\eta_1 = \lim_n \theta_n$, where
$$\theta_n \equiv a^n \xi_n - a^n \eta_{2n} = \eta_1 - a^{-n} \eta_{2n+1}$$
and
$$(a^n \xi_n - a^n \eta_{2n}) \in (\mathcal{F} \vee \mathcal{E})$$
This means that
$$(\mathcal{F} \vee \mathcal{E}) = (\mathcal{G} \vee \mathcal{E})$$
Assuming the modularity equality
$$\mathcal{F} \vee (\mathcal{E} \wedge \mathcal{G}) = (\mathcal{F} \vee \mathcal{E}) \wedge \mathcal{G}$$
and using that $\mathcal{F} \vee \mathcal{E} = \mathcal{G} \vee \mathcal{E}$ we obtain
$$\mathcal{F} = \mathcal{F} \vee 0 = \mathcal{F} \vee (\mathcal{E} \wedge \mathcal{G}) = (\mathcal{F} \vee \mathcal{E}) \wedge \mathcal{G} = (\mathcal{G} \vee \mathcal{E}) \wedge \mathcal{G} = \mathcal{G}$$
That is $\mathcal{F} = \mathcal{G}$, which contradicts $\mathcal{F} \neq \mathcal{G}$. □

Proposition 4.5 $\mathcal{P}(\mathcal{H})$ *is orthomodular, irrespective of the dimension of* $\mathcal{H}$.

Proof: Let $\mathcal{F} \leq \mathcal{G}$. $\mathcal{G}$ is a Hilbert space having $\mathcal{F}$ as a closed linear subspace. $\mathcal{G} \wedge \mathcal{F}^\perp$ is the orthogonal complement of $\mathcal{F}$ in $\mathcal{G}$, i.e.
$$\mathcal{F} \vee (\mathcal{G} \wedge \mathcal{F}^\perp) = \mathcal{G}$$

□

Remark: In their paper in which von Neumann and Birkhoff laid down the foundations of quantum logic, [21] they considered the modularity of the lattice that should be a "quantum logic" an important and necessary property to require. Of course, Birkhoff and von Neumann had been aware that $\mathcal{P}(\mathcal{H})$ is not a modular lattice if $\mathcal{H}$ is the usual (infinite dimensional) Hilbert space of quantum mechanics (the proof of Proposition 4.4 is due to Birkhoff and von Neumann). Consequently, for Birkhoff and von Neumann "quantum logic" was *not* the lattice $\mathcal{P}(\mathcal{H})$. In Section 7 we shall return to the problem what structure (and why) Birkhoff and von Neumann considered as quantum logic.

Proposition 4.6 *Let ϕ be a state on $\mathcal{P}(\mathcal{H})$ (in the sense of Definition 2.6). Then ϕ is a Jauch-Piron state (Definition 3.22).*

Proof: By Gleason's theorem (Proposition 2.4) ϕ is given by a density matrix $\phi(E) = \text{Tr}(wE)$, with a w that can be written in its spectral resolution as $w = \sum_i \lambda_i P_i$ ($\sum_i \lambda_i = 1$). Let A, B be two projections, and assume that $\phi(A) = \phi(B) = 1$. It follows that every projection P_i is contained in both A and in B, so every P_i is contained also in $A \wedge B$, hence $\phi(A \wedge B) = 1$. □

Obviously, the one dimensional subspaces are atoms in $\mathcal{P}(\mathcal{H})$, and it is also clear that every subspace $\mathcal{E}$ is equal to the least upper bound of all subspaces $\mathcal{E}$ majorizes. Furthermore, the following so-called "minimal covering property" holds:

Proposition 4.7 *For any $\mathcal{G} \in \mathcal{P}(\mathcal{H})$ and for any atom $\mathcal{G}_1 \in \mathcal{P}(\mathcal{H})$ such that $\mathcal{G}_1 \wedge \mathcal{G} = 0$ holds we have: $\mathcal{G} \leq (\mathcal{G} \vee \mathcal{G}_1)$ and, if $\mathcal{F} \in \mathcal{P}(\mathcal{H})$ is such that $\mathcal{G} \leq \mathcal{F} \leq (\mathcal{G} \vee \mathcal{G}_1)$, then either $\mathcal{G} = \mathcal{F}$ or $\mathcal{F} = (\mathcal{G} \vee \mathcal{G}_1)$.*

Proof: Let $\mathcal{G}$ be the subspace generated by the elements $\eta_n \in \mathcal{H}$ ($n = 1, 2, \ldots$) and let $\mathcal{G}_1$ be the one-dimensional projection onto the subspace spanned by $\zeta \in \mathcal{H}$. If

$$\mathcal{G} \leq \mathcal{F} \leq (\mathcal{G} \vee \mathcal{G}_1)$$

then any $\eta \in \mathcal{F}$ can be written as

$$\eta = \sum_n \lambda_n \eta_n + \lambda \zeta$$

If $\lambda = 0$ for all $\eta \in \mathcal{F}$, then $\mathcal{F} = \mathcal{G}$; if for some η in $\mathcal{F}$ λ is not equal to zero, then $\mathcal{F} = (\mathcal{G} \vee \mathcal{G}_1)$. □

Proposition 4.8 *Let $\mathcal{H}$ be a Hilbert space. If $\dim \mathcal{H} \geq 3$, then $\mathcal{P}(\mathcal{H})$ does not contain a prime filter (prime ideal).*

Proof: Indirect. Assume that $\mathcal{F} \subset \mathcal{P}(\mathcal{H})$ is a prime filter. Then there exists a projection $A \in \mathcal{P}(\mathcal{H})$ such that $I \neq A \notin \mathcal{F}$. Then $A^\perp$ is a non-zero projection, so there exists a one dimensional projection $P \leq A^\perp$ spanned by a vector $\xi \in A^\perp$. Consider the projection $A \vee (A^\perp - P)$. We claim that

$$(A \vee (A^\perp - P)) \notin \mathcal{F}$$

Assume the contrary. Then either $A \in \mathcal{F}$ or $(A^\perp - P) \in \mathcal{F}$ by the prime filter property of $\mathcal{F}$. Since $A \notin \mathcal{F}$ by assumption, $(A^\perp - P) \in \mathcal{F}$ must be the case. But then by closedness of $\mathcal{F}$ with respect to the meet operation

$$(A \vee (A^\perp - P)) \wedge (A^\perp - P) = A \in \mathcal{F} \tag{4.3}$$

must hold, which again contradicts the assumption $A \notin \mathcal{F}$. So we must have
$$(A \vee (A^\perp - P)) \notin \mathcal{F}$$
and $(A \vee (A^\perp - P))$ is at least a two dimensional projection, since the projection $(A \vee (A^\perp - P))$ and the one dimensional P are orthogonal and span the Hilbert space $\mathcal{H}$ whose dimension is ≥ 3. Hence there exists two non-zero, orthogonal vectors
$$\eta, \theta \in (A \vee (A^\perp - P)) \tag{4.4}$$
and then the two, one dimensional projections P_1, P_2 spanned by the vectors $\xi + \eta$ and $\xi + \theta$ are such that

$$P_1 \wedge P_2 = 0 \tag{4.5}$$
$$(A \vee (A^\perp - P)) \vee P_1 = I \tag{4.6}$$
$$(A \vee (A^\perp - P)) \vee P_2 = I \tag{4.7}$$

Since $(A \vee (A^\perp - P)) \notin \mathcal{F}$, by the prime filter property of $\mathcal{F}$ and by (4.6) and (4.7) we must have $P_1, P_2 \in \mathcal{F}$, but then $P_1 \wedge P_2 = 0 \in \mathcal{F}$ also, which cannot be the case since $\mathcal{F}$ is a proper filter. □

As a consequence of the above proposition and Proposition 3.11 we have

Proposition 4.9 *If the dimension of the Hilbert space $\mathcal{H}$ is greater than, or equal to 3, then there does not exist a lattice homomorphism from $\mathcal{P}(\mathcal{H})$ into the two element Boolean algebra.*

In other words, there does not exist a two-valued evaluation map on a Hilbert lattice $\mathcal{P}(\mathcal{H})$ with $\dim \mathcal{H} \geq 3$.

The above proposition can be proved also by referring to the Gleason's theorem (Proposition 2.4) and to the nonexistence of dispersion-free states on a Hilbert space (Proposition 2.5): Let $\mathcal{H}_n$ be an n-dimensional Hilbert space, and assume that there exists a lattice homomorphism
$$h: \mathcal{P}(\mathcal{H}_n) \to \mathcal{B}_2$$
where we can take $\{0, 1\}$ as the two element Boolean algebra $\mathcal{B}_2$. Then the homomorphism properties of h imply that h has the properties of a state on $\mathcal{P}(\mathcal{H}_n)$ (Definition 2.6). To see this, one has to check the additivity property, which in the finite dimensional case is finite additivity:
$$h(A + B) = h(A) + h(B) \quad \text{for} \quad A \perp B \tag{4.8}$$
And this equality holds because if $A \perp B$, i.e. if $AB = BA = 0$ then
$$0 = h(AB) = h(BA) = h(B)h(A) = h(A)h(B)$$

that is $h(A)$ and $h(B)$ are orthogonal, and so

$$h(A + B) = h(A \vee B) = h(A) \vee h(B) = h(A) + h(B) \qquad (4.9)$$

Since every state on $\mathcal{H}_n$ is of the form $h(A) = Tr(wA)$ by Gleason's theorem, and $A \mapsto Tr(wA)$ cannot be dispersion-free, such a h does not exist on $\mathcal{P}(\mathcal{H}_n)$. If $\mathcal{H}$ is an infinite dimensional Hilbert space, then the lattice $\mathcal{P}(\mathcal{H}_n)$ can be embedded into $\mathcal{P}(\mathcal{H})$ by an embedding $g \colon \mathcal{P}(\mathcal{H}_n) \to \mathcal{P}(\mathcal{H})$, and if $h \colon \mathcal{P}(\mathcal{H}) \to \mathcal{B}_2$ is a Boolean algebra homomorphism, then $g \circ h$ is a Boolean algebra homomorphism from $\mathcal{P}(\mathcal{H}_n)$ into $\mathcal{B}_2$, i.e. $g \circ h$ is a dispersion-free state on $\mathcal{P}(\mathcal{H}_n)$, which is not possible. Consequently, there exist no lattice homomorphism from $\mathcal{P}(\mathcal{H})$ into $\mathcal{B}_2$.

The above argument shows that in fact more follows from Gleason's theorem and from the fact that dispersion-free states do not exist: One can strengthen the Proposition 4.9 by weakening its assumptions. The essential part of the argument showing that $h \colon \mathcal{P}(\mathcal{H}) \to \mathcal{B}_2$ lattice homomorphisms do not exist was to establish that the restriction of h to the projection lattice of a finite dimensional subspace is additive on orthogonal (hence commuting) projections. So if the map $h \colon \mathcal{B}(\mathcal{H}) \to \mathcal{A}$ is a *partial algebra* homomorphism (Definition 3.20) into a *commutative* partial algebra $\mathcal{A}$, then its restriction $h \colon \mathcal{P}(\mathcal{H}) \to \mathcal{B}$ ($\mathcal{B}$ being the Boolean algebra of idempotents of $\mathcal{A}$) has the additivity property (4.8) because h is additive on commuting projections (see eq. (3.31)). Hence, if $g \colon \mathcal{B} \to \mathcal{B}_2$ is a lattice homomorphism with $g(I) = I$, then $g \circ h$ also is additive on orthogonal projections, and so the restriction of $g \circ h$ to the projection lattice of a finite dimensional subspace $\mathcal{H}_n$ ($n \geq 3$) is a dispersion-free state on $\mathcal{H}_n$, which is not possible. So we have

Proposition 4.10 *If* $\dim(\mathcal{H}) \geq 3$, *then there exists no partial algebra homomorphism from* $\mathcal{B}(\mathcal{H})$ *into a commutative algebra; in particular, there exists no partial Boolean algebra homomorphism from* $\mathcal{P}(\mathcal{H})$ *into a Boolean algebra.*

Summing up the properties of $\mathcal{P}(\mathcal{H})$: Let $\mathcal{H}$ be an infinite dimensional complex Hilbert space. Then $\mathcal{P}(\mathcal{H})$ is an atomic, completely atomistic, orthomodular, non-modular, complete lattice having the covering property and not having any prime filter or prime ideal, hence not having any lattice homomorphism into a Boolan algebra. There exists no partial Boolean algebra homomorphism from $\mathcal{P}(\mathcal{H})$ into a Boolean algebra either. If $\mathcal{H}$ is separable, then $\mathcal{P}(\mathcal{H})$ is separable in the sense that any set of mutually orthogonal projections is countable. $\mathcal{P}(\mathcal{H})$ is called the (concrete) quantum logic associated with the quantum system described by the Hilbert space $\mathcal{H}$. More generally, an orthomodular lattice, not necessarily having the form $\mathcal{P}(\mathcal{H})$, is ususally called an *abstract* quantum logic.

4.2. Subspaces and projections

The linear operator A defined on $\mathcal{H}$ is a projection if it is selfadjoint and idempotent (see Definition 2.1). The projections are continuous (in the norm) operators (they are bounded operators), the norm of a projection is equal to 1 (except for the 0 projection). Special projections are the zero 0 and the identity operator I.

There is a one-to-one correspondence between the projections defined on $\mathcal{H}$ and the closed linear subspaces of $\mathcal{H}$: One can identify the projection $E \in \mathcal{P}(\mathcal{H})$ with the *range* of E: $range(E)$ is the linear subspace defined by

$$range(E) = \{\eta \mid E\eta = \eta\}$$

$range(E)$ is a closed linear subspace in $\mathcal{H}$. Conversely, if $\mathcal{G} \subset \mathcal{H}$ is a closed linear subspace, then any $\eta \in \mathcal{H}$ can be written in the form $\eta_1 + \eta_2 = \eta$ such that $\eta_1 \in \mathcal{G}$, $\eta_2 \in \mathcal{G}^\perp$. The operator G defined by $G\eta \equiv \eta_1$ is a projection with $\mathcal{G}$ as $range(G)$. In what follows we shall therefore identify the closed linear subspaces with the projections, denoting by $\mathcal{P}(\mathcal{H})$ both the set of all projections and the set of all closed linear subspaces. If it is necessary to make clear whether an element in $\mathcal{P}(\mathcal{H})$ is viewed as a subspace or as a projection, the convention will be used that the subspaces determined by the projections E, F, G are denoted by the corresponding curly letters $\mathcal{E}, \mathcal{F}, \mathcal{G}$.

Since $\mathcal{P}(\mathcal{H})$ is a subset in the algebra $\mathcal{B}(\mathcal{H})$ of all bounded operators defined on $\mathcal{H}$, there are algebraic operations defined between projections: for every $A, B \in \mathcal{P}(\mathcal{H})$, the $A + B$, AB and λA are all well defined as linear operators, which are, however, not projections in general. In the present subsection we wish to describe the connection between the algebraic and lattice operations in $\mathcal{P}(\mathcal{H})$.

Definition 4.5 The projection F is said to be *smaller* than the projection G (in notation $F \prec G$), if and only if $\langle F\eta, \eta \rangle \leq \langle G\eta, \eta \rangle$ for every $\eta \in \mathcal{H}$.

The next proposition tells us that the relation $\prec$ coincides with the relation of partial ordering defined (Definition 4.1) between the projections as subspaces.

Proposition 4.11 $\mathcal{F} \leq \mathcal{G}$ *if and only if* $F \prec G$.

Proof: Let $\mathcal{F} \leq \mathcal{G}$. Consider the orthogonal decomposition $\eta = \eta^\mathcal{F} + \eta^{\mathcal{F}^\perp}$ Then we have

$$\langle \eta, F\eta \rangle = \langle \eta, F^2\eta \rangle = \langle F\eta, F\eta \rangle = \| \eta^\mathcal{F} \|^2$$

Since $G\eta = \eta$ it holds that

$$\langle \eta, G\eta \rangle = \langle \eta, G^2(\eta^\mathcal{F} + \eta^{\mathcal{F}^\perp}) \rangle = \| \eta^\mathcal{F} + G\eta^{\mathcal{F}^\perp} \|^2 \geq \| \eta^\mathcal{F} \|^2$$

Conversely, if $F \prec G$, then for all $\eta \in \mathcal{H}$ one has

$$\| F\eta \|^2 \leq \| G\eta \|^2$$

If $\eta \in \mathcal{F}$, then

$$\| \eta \| = \| F\eta \| \leq \| G\eta \| \leq \| \eta \|$$

thus $\| G\eta \| = \| \eta \|$, consequently $\eta \in \mathcal{G}$. □

The next proposition characterizes the ordering between projections in terms of the algebraic operations.

Proposition 4.12 $A \leq B$ if and only if $AB = BA = A$

Proof: $A \leq B$ if and only if $BA\eta = A\eta$ for all $\eta \in \mathcal{H}$. But then (and only then)

$$BA = A^*B^* = (BA)^* = A^* = A$$

□

The next proposition links the algebraic operations between projections to the lattice operations.

Proposition 4.13
(i) $A \wedge B = \lim_{n \to \infty}(AB)^n = \lim_{n \to \infty}(BA)^n = \lim_{n \to \infty}(ABA)^n = \lim_{n \to \infty}(BAB)^n$
(ii) $A \vee B = I - \lim_{n \to \infty}\{(I - A)(I - B)\}^n$
(iii) $A^\perp = I - P$

(The limit in (i) and (ii) above, thus also in the proof, is to be understood in the strong sense i.e. $\lim_n X_n = Y$ if and only if $\lim_n X_n\eta = Y\eta$ for every $\eta \in \mathcal{H}$.)

Proof: (i) Put $E_n \equiv (ABA)^n$. Then every E_n is selfadjoint and positive. Positive is also $I - ABA$. We claim that $E_n \geq E_{n+1}$. Indeed, if n is even, $n = 2k$, then

$$\langle \eta, (E_n - E_{n+1})\eta \rangle = \langle \eta, (ABA)^{2k}(I - ABA)\eta \rangle$$
$$= \langle (ABA)^k\eta, (I - ABA)(ABA)^k\eta \rangle \geq 0$$

If $n = 2k + 1$, then

$$\langle \eta, (E_n - E_{n+1})\eta \rangle = \langle \eta, (ABA)^{2k}(ABA - (ABA)^2)\eta \rangle =$$
$$\langle (ABA)^k\eta, (ABA - (ABA)^2)(ABA)^k\eta \rangle \geq 0$$

(since $ABA - (ABA)^2$ is positive). E_n is then a monotone decreasing sequence of positive operators, thus it has a limit E (in the strong topology), which is selfadjoint and positive. We show that $E^2 = E$. Indeed

$$\begin{aligned} E_j E &= E_j \lim_{n\to\infty} E_n \\ = \lim_{n\to\infty} E_j E_n &= \lim_{n\to\infty} E_{j+n} = E \end{aligned}$$

for every j, thus

$$\begin{aligned} E^2 &= (\lim_{n\to\infty} E_n) E \\ = \lim_{n\to\infty} (E_n E) &= \lim_{n\to\infty} E = E \end{aligned}$$

Hence E is a projection. It holds that $A \wedge B \leq E$, since if

$$\eta \in \mathrm{range}(A) \wedge \mathrm{range}(B)$$

then $E\eta = \eta$. Furthermore

$$EA = AE = E$$

and so $E \leq A$ (due to (i)). One still has to see that $E \leq B$. Consider the sequence $F_n \equiv (BAB)^n$. Everything that has been said of E_n, also holds for F_n, thus there exists the projection $F = \lim_{n\to\infty} F_n$ for which $A \wedge B \leq F$ and $F \leq B$. We show that $E = F$. Since

$$(ABA)^j (BAB)^k (ABA)^n = (ABA)^{j+k+n+1}$$

it follows that $EFE = E$, and for similar reasons $FEF = F$. Consequently

$$\begin{aligned} \|E\eta\|^2 = \langle \eta, E\eta \rangle &= \langle \eta, EFE\eta \rangle \\ = \langle E\eta, FE\eta \rangle &= \|FE\eta\|^2 \\ = \|E\eta\|^2 - \|E\eta - FE\eta\|^2 \end{aligned}$$

for every $\eta \in \mathcal{H}$, and so $E = FE$ and $F = EF$, i.e. $E = F$. Thus the third and fourth equality in (i) is proved. To prove the first and second equality in (i) it is enough to see one of them because A and B have a symmetric role in them. We show that $\| (AB)^n \eta - E\eta \|$ tends to zero if $n \to \infty$, where E is the projection constructed above

$$\begin{aligned} & \| (AB)^n \eta - E\eta \| \\ \leq & \| (AB)^n \eta - (ABA)^{n-1} \eta \| + \| (ABA)^{n-1} \eta - E\eta \| \\ = & \| (ABA)^{n-1} (I - B) \eta \| + \| (ABA)^{n-1} \eta - E\eta \| \end{aligned}$$

The second term in the last iquality above tends to zero if $n \to \infty$ by the definition of E; the first term tends to zero because $(I - B)$ is orthogonal to the range of B, thus also to the range of $A \wedge B$. (ii) follows from the properties of orthocomplementation and from (i); (iii) is obvious. □

Proposition 4.14 *Let A and B be projections. Then*
(i) *If $AB = BA$, then $A \wedge B = AB$ and $A \vee B = A + B - AB$*
(ii) *If $A \leq B$, then $B \wedge A^\perp = B - A$*
(iii) *$A \perp B$ if and only if $AB = BA = 0$*

Proof: (i) is an easy consequence of (i) in Proposition 4.13, (ii) and (iii) are obvious. □

The last two propositions give the meet and join of *two* arbitrary projections in terms of the algebraic operations. The meet and join of a countable set of projections can be nicely expressed in algebraic terms only if the projections are special:

Proposition 4.15
(i) *Let A_n, $(n = 1, 2, \ldots)$ be pairwise orthogonal projections.*
   Then $\vee_n A_n = \sum_n A_n$.
(ii) *Let A_n, $(n = 1, 2, \ldots)$ be pairwise commuting projections.*
   Then $\wedge_n A_n = \lim_{n \to \infty} A_1 A_2 \ldots A_n$.

(The limits above are to be understood in the sense of the strong topology.)

Proof: (i) Let $E \equiv \vee_n A_n$. Then $\eta \in \mathcal{H}$ can be decomposed in the form $\eta = \eta_0 + \sum_n \eta_n$, where η_0 is orthogonal to the range of E and $\eta_n \in \text{range} A_n$. One has to show that $\| (E - \sum_{n=1}^{k} A_n)\eta \|$ tends to zero if k tends to infinity.

$$\| (E - \sum_{n=1}^{k} A_n)\eta \|^2 = \| (E - \sum_{n=1}^{k} A_n)(\eta_0 + \sum_j \eta_j) \|$$

$$= \| (E - \sum_{n=1}^{k} A_n)\sum_j \eta_j \| = \| \sum_j (E - \sum_{n=1}^{k} A_n)\eta_j \|^2$$

$$= \| \sum_{j=k+1} \eta_n \|^2 = \sum_{j=k+1} \| \eta_n \|^2$$

The sequence $\sum_{j=k+1} \| \eta_n \|^2$ is convergent (its sum is $\| \eta - \eta_0 \|$), hence $\sum_{j=k+1} \| \eta_n \|^2$ tends to zero if $j \to \infty$.

(ii) Let $E_n \equiv A_1 A_2 \ldots A_n$. By the Proposition 4.13 $E_n = \wedge_n A_n$. Then

$$E_{n-1} \geq E_n \geq 0$$

and the monotone decreasing sequence of operators E_n has a limit E which is selfadjoint. Since

$$E_j E_n = E_n E_j = E_j$$

if $n \geq j$, we have

$$E_j E = E_j \lim_{n \to \infty} E_n$$
$$= \lim_{n \to \infty} E_j E_n = E$$

and similarly $EE_j = E$ for all j. Hence

$$E^2 = E\lim_{n\to\infty} E_n$$
$$= \lim_{n\to\infty} EE_n = E$$

The E is thus a projection. By $EE_j = E$ it holds that $E \leq E_j$ for all j, so

$$E \leq \wedge_n E_n = \wedge_n A_n$$

Obviously $\wedge_n A_n \leq E$, i.e. $E = \wedge_n A_n$. □

A subset in $\mathcal{P}(\mathcal{H})$ is called a *sub-quantum logic* if it is a quantum logic in itself, i.e. if it is closed with respect to the operations $\wedge, \vee$ and $\perp$ defined in $\mathcal{P}(\mathcal{H})$. For instance, if $\mathcal{H}_1 \subseteq \mathcal{H}$ is a closed linear subspace, then $\mathcal{P}(\mathcal{H}_1)$ is a sub-quantum logic. The intersection of sub-quantum logics is again a sub-quantum logic, thus if $\mathcal{P}_0 \subset \mathcal{P}(\mathcal{H})$ is an arbitrary set of projections (subspaces), then there exists the sub-quantum logic generated by $\mathcal{P}_0$, by definition this is the smallest sub-quantum logic containing $\mathcal{P}_0$. A sub-quantum logic $\mathcal{P}_0$ of $\mathcal{P}(\mathcal{H})$ is called *commutative* if $AB = BA$ for all $A, B \in \mathcal{P}_0$.

Proposition 4.16 *The sub-quantum logic $\mathcal{P}_0$ is commutative if and only if $\mathcal{P}_0$ is a distributive lattice.*

Proof: If $\mathcal{P}_0$ is commutative, then by Proposition 4.15 for all $A, B \in \mathcal{P}_0$ it holds that

$$A \wedge B = AB$$
$$A \vee B = A + B - AB$$

hence

$$A \wedge (B \vee C) = A(B \vee C) = A(B + C - BC) = AB + AC - ABC$$

and

$$(A \wedge B) \vee (A \wedge C) = (AB) \vee (AC)$$
$$= AB + AC - ABAC = AB + AC - AABC$$
$$= AB + AC - ABC$$

The dual equality

$$A \vee (B \wedge C) = (A \vee B) \wedge (A \vee C)$$

can be seen to hold in just the same way. If $\mathcal{P}_0$ is distributive, then for all $A, B \in \mathcal{P}_0$ we have

$$B \perp A \wedge (A \wedge B)^\perp$$

since
$$B \perp A \wedge (A \wedge B)^\perp$$
if and only if
$$B \leq (A \wedge (A \wedge B)^\perp)^\perp$$
however, by the distributivity
$$\begin{aligned}(A \wedge (A \wedge B)^\perp)^\perp &= (A \wedge (A^\perp \vee B^\perp))^\perp \\ = ((A \wedge A^\perp) \vee (A \wedge B^\perp))^\perp &= (A \wedge B^\perp)^\perp \\ = A^\perp \vee B \geq B\end{aligned}$$

Since $A \geq A \wedge B$ one has
$$A \wedge (A \wedge B)^\perp = A - (A \wedge B)$$
and because of
$$A = (A \wedge B) + A - (A \wedge B)$$
one can write
$$\begin{aligned}BA = B((A \wedge B) + A - (A \wedge B)) &= B(A \wedge B) + B(A - (A \wedge B)) \\ = B(A \wedge B) + B(A \wedge (A \wedge B)^\perp) &= B(A \wedge B) = A \wedge B \\ AB = ((A \wedge B) + A - (A \wedge B))B &= (A \wedge B)B + (A \wedge (A \wedge B)^\perp)B \\ = (A \wedge B)B &= A \wedge B\end{aligned}$$
(recall that
$$B(A \wedge (A \wedge B)^\perp) = (A \wedge (A \wedge B)^\perp)B = 0$$
if and only if
$$B \perp (A \wedge (A \wedge B)^\perp)$$
□

Proposition 4.17 *The sub-quantum logic $\mathcal{P}_0$ generated by the pairwise commuting projections $\{A_i \mid i = 1, 2 \ldots n \ldots\}$ is distributive.*

Proof: One must show that $\mathcal{P}_0$ is commutative. To this end it suffices to show that there exists a commutative sub-quantum logic containing $\mathcal{P}_0$. Let $\mathcal{M}$ be the von Neumann algebra generated by $\mathcal{P}_0$. $\mathcal{M}$ is the closure in the strong operator topology of the linear sums of projections A_i, and so it is a commutative algebra. Obviously $\mathcal{P}_0$ is part of the sub-quantum logic formed by the set of all projections of $\mathcal{M}$. (Concerning the operator algebraic notions used here see the section 6.1) □

4.3. Bibliographic notes

The lattice $\mathcal{P}(\mathcal{H})$ is described in many quantum logic textbooks, e.g. in [164] and in [18]. For the elementary properties of the Hilbert space see [58], where the example in the Remark after the Definition 4.3 is taken from. The example in the proof of Proposition 4.4 showing that $\mathcal{P}(\mathcal{H})$ is not modular if $\mathcal{H}$ is not finite dimensional is due to Birkhoff and von Neumann [21]. The non-existence of partial algebra homomorphisms from $\mathcal{P}(\mathcal{H})$ into a Boolean algebra was first shown in [85]. The detailed analysis and ramifications of this fact, which is part of what became known as "Kochen-Specker theorem", are discussed at length in [85] and in [134]; we shall briefly return to this issue in the section 9.4. The section 4.2 is based on [98].

CHAPTER 5

Physical theory in semantic approach

By "logic of a physical system" – in broad sense – is meant below the system of logical relations of a set of propositions that are considered meaningful and empirically checkable according to a particular theory that describes the physical system. "Logical relation of propositions" means relations of propositions from the the point of view of "truth", "falsity", "entailment", etc. The analysis of the logic of a physical system can, and will in this chapter, be carried out in the usual terms of object language–metalanguage, syntactic–semantics. The object language is formed by the elementary sentences determined by the physical theory, the semantic notions such as truth, interpretation, evaluation, etc., are determined by the special features of the theory in question.

"Logic of a physical system" in a narrower sense will mean the algebraic structure that represents the equivalence classes of the elementary sentences with respect to the equivalence relation "the sentence A is true if and only if the sentence B is true". It will be seen in Section 5.2 that this algebraic structure is a Boolean algebra in the case of classical mechanics, it is the Tarski-Lindenbaum algebra of the propositional system determined by the elementary sentences of a classical mechanical system. In Section 5.3 the problem is investigated whether/in what sense the lattice of projections $\mathcal{P}(\mathcal{H})$ can be viewed as the "quantum mechanical Tarski-Lindenbaum algebra" of the propositional calculus determined by the elementary sentences of quantum mechanics.

5.1. Physical theory as semi-interpreted language

Let T be a physical theory describing the physical system S. T is conceived as a theory, according to which the system S is in a definite state at each moment and at every moment T allows us to form "elementary sentences" regarding the system. The elementary sentences are empirical statements concerning the values of the observables of S. A typical elementary sentence is

$sent(Q, A) =$ "The value of the observable Q lies in the set A"

where A is a set of real numbers e.g. an interval. Another type of elementary sentence is

$sent(Q, A, r) =$ "The probability that the value of the observable Q lies in the set A is equal to r"

r being a real number between 0 and 1. We shall encounter these latter, probabilistic elementary sentences in connection with quantum mechanics in Section 5.3, but to make things easier to begin with, only the first type of elementary sentences are considered in this section. Let K denote the set of all elementary sentences with respect to a given physical theory T. We wish to define the set $\mathcal{F}$ of *all meaningful* statements determined by T. This set $\mathcal{F}$ will be a set that contains K by definition; however, at this level of generality we do not want to decide whether $\mathcal{F}$ is strictly larger than K. In every concrete application of the present scheme $\mathcal{F}$ will be defined precisely.

Whether the value of an observable quantity Q lies in A depends on which state the system is in. In some states the value of Q lies in A, in some other states it does not. It might also be the case that fixing the state of the system does not determine the truth value of *all* elementary sentences. We shall encounter this situation in quantum mechanics. This relation between the elementary sentences and states of the system can be expressed by a function h from K into the set of subsets of the state space Γ: to every elementary sentence

$$sent(Q, A) \in K$$

we assign a subset

$$h(sent(Q, A)) \subseteq \Gamma$$

of states having the property that in those states the value of Q lies in A. This fact can also be expressed by saying that the states in $h(sent(Q, A))$ *satisfy* (make true) the elementary sentence $sent(Q, A)$. With the help of h we can define further semantic notions as follows.

(i) $sent(Q, A)$ is *true* in state η if the state η of the system is given by an element in: $\eta \in h(sent(Q, A))$;
(ii) $sent(Q, A)$ is *valid* if and only if $h(sent(Q, A)) = \Gamma$;
(iii) $sent(Q, A)$ is a *semantic entailment* of $sent(Q', A')$ if and only if

$$h(sent(Q', A')) \subseteq h(sent(Q, A))$$

equivalently, we can say that $sent(Q', A')$ *implies* $sent(Q, A)$. This relation will also be denoted by

$$sent(Q', A') \models sent(Q, A)$$

Logically speaking, the map h defines the *interpretations* of the subset K of the formulas $\mathcal{F}$ of the language $\mathcal{L}$: an interpretation is given by a state $\varphi \in \Gamma$.

The quadruple $(\mathcal{L}, \mathcal{F}, \Gamma, h)$ is called the *semi-interpreted language* determined by T. It is not only semi-interpreted, but, so to speak, is a semi-language, because neither the syntax of the object language $\mathcal{L}$ nor the semantics of $\mathcal{L}$, as described above is complete, important features are left unspecified. In particular the following are left undefined:

1. It is not determined whether $\mathcal{F}$ contains elements other than the elementary sentences, i.e. it is left open if the set of formulas contains non-atomic formulas, specifically whether it contains the formulas $\sim k$, $k_1 \wedge k_2$ and $k_1 \vee k_2$ $(k_1, k_2 \in K)$.[1]
2. If $\mathcal{F}$ contains elements other than the atomic formulas, the truth conditions for these are not given.
3. The truth condition of the elementary sentences is defined, yet it is not defined when they are false; in particular the Principle that in every interpretation every elementary sentence is either true or false has not been adopted.

Let us consider now the problem of definition of falsity of the elementary sentence $sent(Q, A)$. One can take the position that $sent(Q, A)$ is false if and only if the sentence $sent(Q, A)$ is not true – for whatever reason. In this case "false" and "not true" are not distinguished, and this kind of negation is called the *exclusion negation*. But this is not the only option, since negation is often interpreted as a positive assertion that a particular state of affairs – taken from a certain set of alternatives – is actually the case. If so interpreted, the negation is called *choice negation*. Thus one can define "false" in the following two ways:

EX $sent(Q, A) \in K$ is false in the interpretation φ if and only if

$$\varphi \notin h(sent(Q, A))$$

CH $sent(Q, A) \in K$ is false in the interpretation φ if $\varphi \in \Gamma_{Q,A}$ for some designated set $\Gamma_{Q,A} \subseteq \Gamma$ of states, where the set $\Gamma_{Q,A}$ can depend on Q, A but $\Gamma_{Q,A} \neq \Gamma \setminus h(sent(Q, A))$.

One can now ask the question: Assuming that the elementary sentence $k \in K$ is false in the sense of **EX**, is there a $k' \in K$ which is true if and only if k is false? In other words we ask if there exists a $k' \in K$ such that $h(k') = \Gamma \setminus h(k)$. Obviously, there is no apriori reason why such a $k' \in K$ should exist, since K is determined by a particular physical theory T, and, in general, there is no reason why, according to T, an observable Q' should exist for which it holds that

$$h(sent(Q', A')) = \Gamma \setminus h(sent(Q, A)) \tag{5.1}$$

---
[1] If it is not important to specify Q and A in an elementary sentence, the sentences (elements of K) will be denoted by k, k_i etc.

for some A'. If there is an $sent(Q', A')$ in K such that (5.1) holds we say that K is *closed* or *complete* with respect to the **EX** negation. But, even if the required k' does not exist as an element of K, if h can be extended from K to $\mathcal{F}$, then there might exist an element α of $\mathcal{F}$ such that

$$h(\alpha) = \Gamma \setminus h(k)$$

and if this is the case, then we say that $\mathcal{F}$ (or that the language $\mathcal{L}$) is closed (or complete) with respect to the **EX** negation. So the answer to the question whether a certain language is closed (complete) with respect to the **EX** negation depends partly on T, and partly on how the set of (non-atomic) formulas is defined. Similarly, one can define the closedness (completeness) of $\mathcal{F}$ with respect to the **CH** negation.

Let $h(k_1), h(k_2) \subseteq \Gamma$, then one can form the intersection

$$h(k_1) \cap h(k_2) \subseteq \Gamma$$

and one can ask if for any two $k_1, k_2 \in K$ there exists a $k \in K$ or $\alpha \in \mathcal{F}$ such that

$$h(k) = h(k_1) \cap h(k_2) \qquad (5.2)$$
$$h(\alpha) = h(k_1) \cap h(k_2) \qquad (5.3)$$

Again, the answer is not apriori given but depends on both T and $\mathcal{F}$. If yes, we say that K (respectively $\mathcal{F}$) is *closed (or complete)* with respect to the conjunction defined by the set theoretical intersection. Also, one can define the completeness of K and $\mathcal{F}$ with respect to the disjunction

$$h(k_1) \cup h(k_2) \subseteq \Gamma$$

or with respect to any other semantically defined logical operation.

To illustrate the introduced notions, the logic of classical mechanics is described next.

5.2. The logic of classical mechanics

Consider the classical mechanical system S of a single point particle. The phase space of S is $\Gamma = \mathbb{R}^6$, the state of the system is given by (q, p) where q, p are the values of the position and momentum observables of the particle. The set of all observables is assumed to be the set $L^\infty(\mathbb{R}^6, \mu)$ of all Lebesgue measurable, essentially bounded real functions on $\mathbb{R}^6$. The typical elementary sentence in this theory is the

$$sent(g, A) = \text{"the value of } g \text{ lies in the set } A\text{"}$$

where $g \in L^\infty(\mathbb{R}^6, \mu)$ and it is assumed that A is from the Boolean algebra $G(\mathbb{R})$ of real, Lebesgue measurable sets. Let K^C denote the set of all elementary sentences. The set $\mathcal{F}^C$ of formulas is defined by induction as follows:

(i) the elements of K^C are formulas;
(ii) if α is a formula, then $(\sim \alpha)$ is formula;
(iii) if α and β are formulas, then $(\alpha \wedge \beta)$ and $(\alpha \vee \beta)$ are formulas.

According to classical mechanics the elementary sentence $sent(g, A)$ is true if the state of the particle lies in the subset $g^{-1}(A) \subseteq \mathbb{R}^6$ of the phase space. Thus the map h (cf. previous section) is defined by

$$h(sent(g, A)) = g^{-1}(A) \tag{5.4}$$

An interpretation of $\mathcal{F}^C$ is given by a state φ of the system S. The truth values of the formulas in the interpretation φ are defined with the help of h as follows:

(i) $k \in K^C$ is true in φ if and only if $\varphi \in h(k)$. If $\varphi \notin h(k)$ then k is false;
(ii) $(\sim \alpha)$ is true in φ if and only if α is false in φ;
(iii) $(\alpha \wedge \beta)$ is true in φ if and only if both α and β is true in φ; $(\alpha \vee \beta)$ is true in ϕ if either α or β is true in ϕ.

(ii) shows that the negation is defined as exclusion negation.

Now we introduce a relation $\Rightarrow$ in $\mathcal{L}$ as follows:

$\alpha \Rightarrow \beta$ iff for every state ϕ " if α is true in ϕ then β is true in ϕ"

This relation $\Rightarrow$ is obviously reflexive and transitive but it is neither symmetric nor antisymmetric, since it can happen that $sent(g, A)$ is true if and only if $sent(f, B)$ is true without being the case that $g = f$.[2] However, one obtains a reflexive, transitive and symmetric (equivalence) relation $\leftrightarrow$ from $\Rightarrow$, if one identifies the formulas that can not be distinguished with respect to their truth value:

$$\alpha \leftrightarrow \beta \text{ if and only if } \alpha \Rightarrow \beta \text{ and } \beta \Rightarrow \alpha$$

Let $\mathcal{F}^C_\leftrightarrow$ denote the equivalence classes of $\mathcal{F}^C$ with respect to the equivalence relation $\leftrightarrow$. If $\alpha \in \mathcal{F}^C$, then the equivalence class of α is denoted by $|\alpha|$. The operations $\sim$ and $\wedge$ in $\mathcal{F}^C$ can be carried over to $\mathcal{F}^C_\leftrightarrow$ by defining:

$$\sim |\alpha| \equiv |\sim \alpha| \text{ and } |\alpha| \wedge |\beta| \equiv |\alpha \wedge \beta|$$

---

[2]Take two functions f and g such that $f(q_0, p_0) = g(q_0, p_0)$ for some q_0, p_0 but $f(q, p) \neq g(q, p)$ for all $(q, p) \neq (q_0, p_0)$ then $sent(g, \{(q_0, p_0)\})$ is true if and only if $sent(f, \{(q_0, p_0)\})$ is true but $sent(g, \{(q_0, p_0)\}) \neq sent(f, \{(q_0, p_0)\})$.

and the relation $\Rightarrow$ defined by

$$|\alpha| \Rightarrow |\beta| \text{ if and only if } \alpha \Rightarrow \beta$$

becomes a partial ordering on $\mathcal{F}_{\leftrightarrow}^C$. (One should prove at this point that these definitions are meaningful, i.e. one should prove that if

$$|\alpha_1| = |\alpha_2| \text{ and } |\beta_1| = |\beta_2|$$

then

$$|\sim \alpha_1| = |\sim \alpha_2| \text{ and } |\alpha_1 \wedge \beta_1| = |\alpha_2 \wedge \beta_2|$$

Rather than writing down explicitly these obvious but tedious checks, we refer to the demonstration after (5.6)-(5.8) where such a check is shown in connection with the quantum analog of the present situation.)

The structure $(\mathcal{F}_{\leftrightarrow}^C, \sim, \wedge)$ is known in logic as the *Tarski-Lindenbaum algebra* determined by $\mathcal{F}^C$. (That this is an algebra indeed, will be seen below.)

It is clear that

$$K_{\leftrightarrow}^C \subseteq \mathcal{F}_{\leftrightarrow}^C$$

We wish to show that

$$\mathcal{F}_{\leftrightarrow}^C \subseteq K_{\leftrightarrow}^C$$

To see this it is enough to show that if

$$k, k_1, k_2 \in K^C$$

then

$$|\sim k| \in K_{\leftrightarrow}^C \text{ and } |k_1 \wedge k_2| \in K_{\leftrightarrow}^C$$

Let $k = sent(f, A) \in K^C$. Consider $(\sim k)$. The formula $(\sim k)$ is true in ϕ if and only if

$$\varphi \in (\mathbb{R}^6 \setminus h(k)) = \mathbb{R}^6 \setminus f^{-1}(A)$$

The set $f^{-1}(A)$ is Lebesgue measurable in $\mathbb{R}^6$, thus $\mathbb{R}^6 \setminus f^{-1}(A)$ also is Lebesgue measurable, consequently the characteristic (indicator) function $I_{\mathbb{R}^6 \setminus f^{-1}(A)}$ belongs to $L^\infty(\mathbb{R}^6, \mu)$, and so it is an observable. Therefore

$$sent(I_{\mathbb{R}^6 \setminus f^{-1}(A)}, \{1\}) \in K^C$$

and $sent(I_{\mathbb{R}^6 \setminus f^{-1}(A)}, \{1\})$ is true in ϕ if and only if $(\sim sent(f, A))$ is true in ϕ. This means that K^C is closed with respect to the exclusion negation, furthermore, if $k \in K^C$, then $|(\sim k)| \in K_{\leftrightarrow}^C$. Similarly, if $sent(f, A)$ and $sent(g, B)$ are elementary sentences in K^C, then

$$sent(I_{f^{-1}(A) \cap g^{-1}(B)}, \{1\}) \in K^C$$

and $sent(I_{f^{-1}(A) \cap g^{-1}(B)}, \{1\})$ is true in ϕ if and only if

$$sent(f, A) \wedge sent(g, B)$$

is true in ϕ. This means that K^C is closed with respect to the conjunction defined by the set theoretical intersection and that if $k_1, k_2 \in K^C$, then $|k_1 \wedge k_2| \in K^C_{\leftrightarrow}$. Obviously, the map h can be "extended" from K^C to a function $|h|$ defined on $K^C_{\leftrightarrow}$ by

$$|h|(|k|) \equiv h(k) \tag{5.5}$$

and the considerations leading to the closedness of $\mathcal{F}^C$ with respect to the classical logical operations show that the extension $|h|$ has these properties:

$$\begin{aligned}
|h|(\sim |k|) &= \Gamma \setminus |h|(|k|) \\
|h|(|k_1| \wedge |k_2|) &= |h|(|k_1|) \cap |h|(|k_2|) \\
|h|(|k_1| \vee |k_2|) &= |h|(|k_1|) \cup |h|(|k_2|)
\end{aligned}$$

That is the extension of $|h|$ becomes a Boolean algebra isomorphism between $K^C_{\leftrightarrow}$ and the Boolean algebra $G(\mathbb{R}^6)$ of Lebesgue measurable sets in the phase space.

We sum up the facts on the logic of a classical mechanical system in the following propositions:

1. The set of elementary sentences in the language determined by a classical mechanical system is closed with respect to the classical logical operations of exclusion negation and "and";
2. the meaningful propositions of the language of a classical mechanical system, i.e. the Tarski-Lindenbaum algebra determined by the rules of the classical propositional calculus is a Boolean algebra.

In short, the logic of a classical mechanical system is a classical propositional logic.

The above proposition remains valid also if one makes the more realistic assumption that two Lebesgue measurable sets A, B whose difference

$$(A \cup B) \setminus (A \cap B)$$

has Lebesgue measure zero cannot be distinguished physically. The set of zero Lebesgue measure sets is an ideal in $G(\mathbb{R})$ and factorizing $G(\mathbb{R})$ by this ideal we obtain a Boolean algebra. Replacing $G(\mathbb{R})$ by this (non-atomic) Boolean algebra, all the considerations leading to the above statements remain valid.

5.3. Hilbert lattice as logic

The aim of this section is to clarify whether/in what sense the projection lattice $\mathcal{P}(\mathcal{H})$ can be viewed as the logic of a quantum mechanical system. Recall that by the logic of a physical system is meant the algebraic structure that represents the empirically checkable propositions and their logical relations from the point of view of truth, falsity, etc. It was seen in the previous Section 5.2 that the logic of classical mechanics, the Tarski-Lindenbaum algebra of the elementary propositions, is a Boolean algebra. In this section the problem is investigated whether/in what sense $\mathcal{P}(\mathcal{H})$ can be considered as the Tarski-Lindenbaum "algebra" of the elementary propositions in quantum mechanics. It will be seen that $\mathcal{P}(\mathcal{H})$ can be meaningfully considered as the representative of the logic of a quantum system in a sense very similar to classical mechanics and its logic. There are very important and characteristic deviations from the classical case, however. The differences are due to two characteristic features of quantum mechanics, namely, first, that quantum mechanics is a probabilistic theory, and, second, that quantum probability is non-commutative.

The probabilistic character of quantum mechanics comes to light already in the way the elementary sentences are defined, or rather, in how they can *not* be defined: If one takes the probabilistic character of quantum mechanics seriously, then one must say that quantum mechanics *does not* give the truth conditions of sentences like $sent(Q,d) =$ "The value of the observable Q lies in the set d". The typical, empirically verifiable elementary sentences in quantum mechanics are

$sent(Q,d,r) \equiv$ "the probability that the value of the observable Q lies in d is equal to r"

r being a real number between zero and one. Let us now limit ourselves only to those probabilistic elementary sentences in which $r = 1$, i.e. let us single out the sentences

$sent(Q,d,1) =$ "The probability that the value of Q lies in d is equal to 1"

where Q is an observable quantity, identified with a selfadjoint operator defined on the Hilbert space $\mathcal{H}$. Let the set of all such elementary sentences be K^q. Let us assume, furthermore, that the sets d that can occur in the elementary sentences $sent(Q,d,1)$ cannot be arbitrary but belong to the Boolean algebra of all Borel measurable real sets.

The set of meaningful sentences (the set of formulas) $\mathcal{F}^q$ are defined in complete analogy with the classical case (cf. Section 5.2):

(i) the elements of K^q are formulas;
(ii) if α is a formula, then $(\sim \alpha)$ is a formula;
(iii) if α and β are formulas, then $(\alpha \wedge \beta)$ and $(\alpha \vee \beta)$ are formulas.

When is the elementary sentence $sent(Q, d, 1)$ true in a state ξ of the system? According to quantum mechanics it is true if the state vector $\xi \in \mathcal{H}$ of the system lies in the closed linear subspace determined by the spectral projection $\mathrm{P}^Q(d)$ of the observable Q. (More precisely, if ξ is in the unit sphere in the subspace $\mathrm{P}^Q(d)$. Since the unit sphere in any closed linear subspace is determined by the subspace and vice versa, there is no danger if one identifies the space of pure states of the quantum system with $\mathcal{H}$, rather than saying the states are elements in the unit sphere in $\mathcal{H}$.) Thus the map h in Section 5.1 is now defined by

$$h(sent(Q, A)) \equiv h^q((sent(Q, d, 1)) \equiv \mathrm{P}^Q(d) \in \mathcal{P}(\mathcal{H})$$

(Compare this definition of h^q with the Definition (5.4)!)

Following the procedure describing the logic of a classical system, the next step is to define the interpretations of $\mathcal{F}^q$. An interpretation is given by a quantum state ξ and the truth values of the formulas are defined with the help of h^q as follows:

(i) $sent(Q, d, 1) \in K^q$ is true in ξ if and only if
$$\xi \in h^q(sent(Q, d, 1)) = \mathrm{P}^Q(d).$$
If
$$\xi \in h^q(sent(Q, d, 1))^\perp = \mathrm{P}^Q(d)^\perp$$
then (and only then) $sent(Q, d, 1)$ is false;

(ii) $(\sim \alpha)$ is true in ξ if and only if α is false in ξ; $(\sim \alpha)$ is false in ξ if and only if α is true. Let $[\alpha]$ denote the set of states (interpretations) that make α true. Then the truth conditions of $(\sim \alpha)$ can be formulated as follows: $(\sim \alpha)$ is true in ξ if and only if $\xi \in [\alpha]^\perp$; $(\sim \alpha)$ is false in ξ if and only if $\xi \in [\alpha]$.

(iii) $(\alpha \wedge \beta)$ is true in ξ if and only if α is true in ξ and β is true in ξ; that is if and only if $\xi \in [\alpha] \wedge [\beta]$
$(\alpha \wedge \beta)$ is false in ξ if and only if

$$\xi \in \Big([\alpha] \cap [\beta]\Big)^\perp$$

$(\alpha \vee \beta)$ is true in ξ if and only if $\sim (\sim \alpha \wedge \sim \beta)$ is true in ξ;
$(\alpha \vee \beta)$ is false in ξ if and only if $\sim (\sim \alpha \wedge \sim \beta)$ is false in ξ.

The above definition differs from the corresponding classsical definition in several important respects. The first of these is that the negation above is a choice negation: $\sim \alpha$ is defined to be true if and only if α is false, which, in case of an atomic formula (elementary sentence) $\alpha = sent(Q, d, 1)$, holds if and only if another well-defined condition is met, namely that

$$\xi \in h^q(sent(Q, d, 1))^\perp = \mathrm{P}^Q(d))^\perp$$

Unlike in the logic of classical mechanics, it is thus *not* the case now that if α is not true then $\sim \alpha$ is true. It may very well be that given a state ξ, α is not true and $\sim \alpha$ is not true either. This is due to the fact that it is not true that given two projections $E, E^{\perp}$, any ξ is either in E or in $E^{\perp}$, a consequence of the non-commutative character of quantum probability. Therefore, and this is the second major deviation in the above definition of the truth values of the meaningful quantum sentences from the classical definition, the above definition of ξ as an interpretation of $\mathcal{F}^q$ is incomplete in the sense that it is not true that given ξ, every formula $\alpha \in \mathcal{F}^q$ is either true or false. Given ξ, there exist formulas which are neither true nor false. So one must add to the definition of interpretation given by [i]-[iii] the following:

(iv) If a formula is neither true nor false in ξ, then its truth value is defined to be indeterminate.

The introduction of a "third truth value" (indeterminate) maybe thought unacceptable and we may want to try to define the negation as exclusion negation to save the bivalence of truth values. The trouble with this is that the quantum propositional calculus is complete with respect to the choice negation (this will be seen below) but it is not complete with respect to the exclusion negation. This seems to be more troublesome than the presence of indeterminate truth values: completeness of the quantum propositional calculus with respect to the negation (and the other logical connectives) should hold if the projection lattice is to represent the logic of the system in any reasonable sense in this semantic approach.

We introduce now a relation $\Rightarrow$ in $\mathcal{F}^q$ just as in the classical case:

$\alpha \Rightarrow \beta$ iff for every state ξ " if α is true in ξ then β is true in ξ"

This relation $\Rightarrow$ is reflexive and transitive but neither symmetric (obviously) nor antisymmetric: clearly the condition

$$h^q(sent(Q,d,1)) = \mathrm{P}^Q(d) = \mathrm{P}^R(d) = h^q(sent(R,d,1))$$

i.e. the equality of one of the spectral projections of the observables Q and R, does not imply $Q = R$. One obtains a reflexive, transitive and symmetric relation from $\Rightarrow$ by identifying the formulas that can not be distinguished with respect to their truth value:

$$\alpha \leftrightarrow \beta \text{ if and only if } \alpha \Rightarrow \beta \text{ and } \beta \Rightarrow \alpha$$

Let $\mathcal{F}^q_{\leftrightarrow}$ be the set of equivalence classes of $\mathcal{F}^q$ with respect to $\leftrightarrow$. If $\alpha \in \mathcal{F}^q$, then the equivalence class of α is denoted by $|\alpha|$. The operations $\sim, \wedge$ in

$\mathcal{F}^q$ can be carried over to $\mathcal{F}^q_\leftrightarrow$ by defining

$$\sim |\alpha| \equiv |\sim \alpha| \qquad (5.6)$$
$$|\alpha| \wedge |\beta| \equiv |\alpha \wedge \beta| \qquad (5.7)$$
$$|\alpha| \vee |\beta| \equiv |\alpha \vee \beta| \qquad (5.8)$$

We must show that the definitions are meaningfull, i.e. we must show that if

$$|\alpha_1| = |\alpha_2| \qquad (5.9)$$
$$|\beta_1| = |\beta_2| \qquad (5.10)$$

then

$$|\sim \alpha_1| = |\sim \alpha_2| \qquad (5.11)$$
$$|\alpha_1 \wedge \beta_1| = |\alpha_2 \wedge \beta_2| \qquad (5.12)$$
$$|\alpha_1 \vee \beta_1| = |\alpha_2 \vee \beta_2| \qquad (5.13)$$

These are routine checks, we only show (5.13).

By the definition of the equivalence classes $|\alpha|$

$$|\alpha_1 \vee \beta_1| = |\alpha_2 \vee \beta_2|$$

if and only if

$$[\alpha_1 \vee \beta_1] = [\alpha_2 \vee \beta_2]$$

if and only if (by the definition of the truth of $(\alpha \vee \beta)$)

$$[\sim (\sim \alpha_1 \wedge \sim \beta_1)] = [\sim (\sim \alpha_2 \wedge \sim \beta_2)]$$

if and only if (by the definition of truth of $\sim \alpha$)

$$[\sim \alpha_1 \wedge \sim \beta_1]^\perp = [\sim \alpha_2 \wedge \sim \beta_2]^\perp$$

if and only if (by the properties of orthocomplementum in Hilbert space)

$$[\sim \alpha_1 \wedge \sim \beta_1] = [\sim \alpha_2 \wedge \sim \beta_2]$$

if and only if (by the definition of truth of $\alpha \wedge \beta$)

$$[\sim \alpha_1] \cap [\sim \beta_1] = [\sim \alpha_2] \cap [\sim \beta_2]$$

if and only if (by the definition of truth of $\sim \alpha$)

$$[\alpha_1]^\perp \cap [\beta_1]^\perp = [\alpha_2]^\perp \cap [\beta_2]^\perp$$

but this is true because (5.9)-(5.10) means

$$[\alpha_1] = [\alpha_2] \quad \text{and} \quad [\beta_1] = [\beta_2]$$

The relation $\Rightarrow$ defined by

$$|\alpha| \Rightarrow |\beta| \text{ if and only if } \alpha \Rightarrow \beta$$

becomes a partial ordering on $\mathcal{F}^q_{\leftrightarrow}$.

Clearly

$$K^q_{\leftrightarrow} \subseteq \mathcal{F}^q_{\leftrightarrow}$$

We show that

$$\mathcal{F}^q_{\leftrightarrow} \subseteq K^q_{\leftrightarrow}$$

To see this it is sufficient to show that if $k, k_1, k_2 \in K^q$, then

$$|\sim k| \in K^q_{\leftrightarrow} \text{ and } |k_1 \wedge k_2| \in K^q_{\leftrightarrow}$$

Let

$$k = sent(Q, d, 1) \in K^q$$

and consider $(\sim k)$. $(\sim k)$ is true in ξ if and only if

$$\xi \in P^Q(d)^\perp$$

Since $P^Q(d)^\perp$ is a projection, it is an observable, thus there is an elementary sentence $sent(Q', d', 1)$, namely the sentence

$$sent(Q', d', 1) = sent(P^Q(d)^\perp, \{1\}, 1)$$

for which it holds that

$$h^q(sent(Q', d', 1)) = P^Q(d)^\perp$$

That is, the elementary sentence

$$sent(Q', d', 1) = sent(P^Q(d)^\perp, \{1\}, 1)$$

is true in ξ if and only if $(\sim sent(Q, d, 1))$ is true in ξ. This means that K^q is closed with respect to the particular choice negation defined for the quantum $\mathcal{F}^q$; furthermore, if $k \in K^q$, then $|(\sim k)| \in K^q_{\leftrightarrow}$. Similarly, if $sent(Q, d, 1)$ and $sent(R, s, 1)$ are elementary sentences in K^q, then

$$sent(Q, d, 1) \wedge sent(R, s, 1)$$

is true in ξ if and only if

$$\xi \in (\mathrm{P}^Q(d) \cap \mathrm{P}^R(s)) = (\mathrm{P}^Q(d) \wedge \mathrm{P}^R(s))$$

Since $(\mathrm{P}^Q(d) \wedge \mathrm{P}^R(s))$ is an observable, the

$$sent((\mathrm{P}^Q(d) \wedge \mathrm{P}^R(s)), \{1\}, 1)$$

is an elementary sentence, which is true in ξ if and only if

$$sent(Q, d, 1) \cap sent(R, s, 1)$$

is true in ξ. This means that K^q is complete with respect to the conjunction defined by the meet of two projections and that, if $k_1, k_2 \in K^q$, then

$$|k_1 \wedge k_2| \in K^q_{\hookrightarrow}$$

Similarly, we can show that $\mathcal{F}^q$ is complete with respect to the "or" and that $(k_1 \vee k_2)$ is represented in $\mathcal{P}(\mathcal{H})$ by $(A_1 \vee A_2)$ if k_1 and k_2 are represented by the projections A_1, A_2.

Obviously, h^q can be "extended" from K^q to $K^q_{\hookrightarrow}$ just as in the case of the logic of a classical system:

$$|h^q|(|k|) \equiv h^q(k) \tag{5.14}$$

and the considerations leading to the closedness of $\mathcal{F}^q$ with respect to the logical operations show that the extension $|h^q|$ is a bijection between $K^q_{\hookrightarrow} = \mathcal{F}^q_{\hookrightarrow}$ and $\mathcal{P}(\mathcal{H})$, furthermore, $|h^q|$ has these properties:

$$\begin{aligned} |h^q|(\sim |k|) &= (|h^q|(|k|))^\perp \\ |h^q|(|k_1| \wedge |k_2|) &= |h^q|(|k_1|) \wedge |h^q|(|k_2|) \\ |h^q|(|k_1| \vee |k_2|) &= |h^q|(|k_1|) \vee |h^q|(|k_2|) \end{aligned}$$

That is the extension $|h^q|$ becomes an isomorphism between $K^q_{\hookrightarrow}$ and $\mathcal{P}(\mathcal{H})$ as an orthocomplemented lattice.

Thus we see that $\mathcal{P}(\mathcal{H})$ represents the set of equivalence classes *of a certain type of* elementary sentences of quantum mechanics with respect to the equivalence relation "k_1 is true if and only if k_2 is true", similarly to the case of classical mechanics. But now $K^q_{\hookrightarrow}$ is, of course, *not* a Boolean algebra but a more general orthomodular lattice. The non-classical character of the logic of a quantum system manifests in several (non-independent) features:

1. The elementary sentences in quantum mechanics are probabilistic, and not even every probabilistic elementary sentence is represented by the elements of the projection lattice: sentences asserting non-trivial probabilities are excluded;

2. The interpretation of the quantum propositional system is not bivalent;
3. The quantum negation is not the exclusion negation but a choice negation;
4. The distributivity law is not valid;
5. The "k_1 or k_2"= $k_1 \vee k_2$ can be true without k_1 or k_2 being true.

One may say that these non-classical features of the quantum logic are counterintuitive: they differ from both the everyday usage of the connectives "and", "not" and "or" and from the properties of these logical operations in the context of classial mechanics. So one might say that because of these non-classical features $\mathcal{P}(\mathcal{H})$ cannot be considered the logic of the quantum system. Certainly one can have a notion of logic, which is closer to our everyday intuition and with respect to which quantum logic does not qualify as logic. Logic can, however be understood differently as a kind of logic whose "... nature is determined by quasi-physical and technical reasoning, different from the introspective and philosophical considerations which have had to guide logicians hitherto." [21] ([178] p.119.) Accepting this view of logic, it is perfectly reasonable to say that $\mathcal{P}(\mathcal{H})$ does represent the logical relations between certain propositions that are considered empirically testable in quantum mechanics.

While the Hilbert lattice $\mathcal{P}(\mathcal{H})$ represents the logic of *probability one* quantum propositions, it should be emphasized that under the present interpretation $\mathcal{P}(\mathcal{H})$ by no means represents the logic of *all* empirically meaningful and verifiable quantum propositions. Take an elementary sentence $sent(Q,d,r)$ ="The probability that Q takes its value in d is equal to r", where $0 < r < 1$. The states ξ that make this sentence true are the vectors for which $\langle \xi, \mathrm{P}^Q(d)\xi \rangle = r$, that is these states lie on a "cone" around the closed linear subspace $\mathrm{P}^Q(d)$, a cone with the angle $2\arccos(r)$. It is clear then that the set of these states is not even a linear subspace in $\mathcal{H}$, and the propositional system of these sentences can certainly not be $\mathcal{P}(\mathcal{H})$. This leads to the question as to *what* is then the propositional system determined by the more general set of probabilistic elementary sentences, where the probabilities are not trivial? We know of no work that treats this problem.

5.4. Bibliographic notes

The notion of "semi interpreted language" determined by a physical theory is due to Van Fraassen [160], he himself refers to Beth concerning the origin of the notion (see the references in [160]). Van Fraassen is also a defender of the semantic approach to physical theories in general (as opposed to the purely formal, syntactic reconstructions), see [163]. It should be noted that it is a controversial issue whether/in what sense $\mathcal{P}(\mathcal{H})$ can be considered as logic; there is no consensus among philosophers of science – or among physicists for that matter – regarding this question. Jauch, as mentioned in the Introduction already, interprets the Hilbert lattice as an empirically given propositional system, and opposes the view that it is "logic" [75]. Putnam, on the other hand goes as far as claiming that logic is empirical and we must accept the highly non-classical logic of quantum mechanics on pain of facing contradictions [117].

CHAPTER 6

Von Neumann lattices

In this chapter the basic properties of von Neumann lattices are described. The von Neumann lattices are the projection lattices of von Neumann algebras, and the von Neumann algebras are those ∗-subalgebras of $\mathcal{B}(\mathcal{H})$ which are closed with respect to certain topologies weaker than the uniform topology in $\mathcal{B}(\mathcal{H})$. Section 6.1 gives the definition of a von Neumann algebra together with the central theorem, von Neumann's double commutant theorem (Proposition 6.1), which characterizes the closedness of ∗-subalgebras of $\mathcal{B}(\mathcal{H})$ (and thereby the von Neumann algebras) purely in algebraic terms. In the Section 6.2 it is proved that the set of projections of a von Neumann algebra is a complete orthomodular lattice (Proposition 6.3). This section also describes briefly the (by now) classic dimension theory of projections of a von Neumann algebra, which is intimately related to the classification theory of von Neumann algebras. Particular attention will be paid to the *finite* von Neumann algebras (which are *not* the same as the *finite dimensional* algebras – although finite diensional algebras are finite). The surprizing fact about a finite algebra is that its projection lattice is not only orthomodular but also modular (Proposition 6.14). The other important proposition in this section is Proposition 6.5, which gives equivalent characterization of the finiteness of an algebra in terms of existence of traces on the algebra. The existence of a (unique) trace on a finite (factor) von Neumann algebra becomes important when it is asked why Birkhoff and von Neumann considered the projection lattices of finite von Neumann algebras as the "proper" quantum logic. The Birkhoff-von Neumann's concept of quantum logic will be discussed in Chapter 7. We shall extensively use the algebraic notions introduced in this chapter in the remaining part of the book.

6.1. Von Neumann algebras

Recall that the linear operator Q defined on the Hilbert space $\mathcal{H}$ is called *bounded* (or continuous) if

$$\sup_{\|\eta\|\leq 1} \| A\eta \| < \infty$$

The domain of definition of a bounded operator is always all of $\mathcal{H}$. $\mathcal{B}(\mathcal{H})$ denotes the set of all bounded operators on $\mathcal{H}$.

$\mathcal{B}(\mathcal{H})$ itself is a linear space over the complex numbers with respect to the sum operation

$$(Q + R)(\eta) \equiv Q\eta + R\eta$$

and multiplication by scalars

$$(Q, \lambda) \mapsto \lambda Q$$

Furthermore, the formula

$$\| Q \| \equiv \sup_{\|\eta\| \leq 1} \| Q\eta \| < \infty$$

defines a norm on $\mathcal{B}(\mathcal{H})$, with respect to which $\mathcal{B}(\mathcal{H})$ becomes a Banach space (i.e. $\mathcal{B}(\mathcal{H})$ is complete in this norm). The topology determined by the operator norm is also referred to as the *uniform* topology in $\mathcal{B}(\mathcal{H})$.

$\mathcal{B}(\mathcal{H})$ also is an algebra if the product is defined by the composition:

$$(QR)(\eta) \equiv Q(R(\eta))$$

What is more, $\mathcal{B}(\mathcal{H})$ is a *Banach algebra*, i.e. it holds that

$$\| QR \| \leq \| Q \| \| R \|$$

(the product is continuous in the norm).

Q^* is the *adjoint* of Q if

$$\langle \eta, Q\zeta \rangle = \langle Q^*\eta, \zeta \rangle$$

for all $\eta, \zeta \in \mathcal{H}$. The map

$$\mathcal{B}(\mathcal{H}) \ni Q \mapsto Q^* \in \mathcal{B}(\mathcal{H})$$

is conjugate linear and has the properties
(i) $(Q^*)^* = Q$,
(ii) $(QR)^* = R^*Q^*$,
(iii) $\| Q^* \| = \| Q \|$,

Definition 6.1 A Banach algebra in which a map $*$ having the above properties (i)-(iii) is called *involutive Banach algebra*.

The following so-called C^*-property also holds in $\mathcal{B}(\mathcal{H})$

(iv) C^*-property: $\quad \| Q^*Q \| = \| Q \|^2 \quad$ for every $Q \in \mathcal{B}(\mathcal{H})$

Definition 6.2 A C^*-*algebra* is an involutive Banach algebra having the C^*-property.

Every norm closed ∗-subalgebra of $\mathcal{B}(\mathcal{H})$ is a C^*-algebra, and the basic Gelfand-Naimark-Segal theorem asserts that given a C^*-algebra $\mathcal{A}$ there exists a ∗-algebra isomorphism (also called: representation) π from $\mathcal{A}$ into the set of all bounded operators $\mathcal{B}(\mathcal{H})$ over some Hilbert space $\mathcal{H}$. The theorem is also known as GNS-construction, since the isomorphism can be constructed from states in the following way. Given a state ϕ (i.e. a positive: $\phi(X^*X) > 0$ for every X; linear: $\phi(\lambda_1 X_1 + \lambda_2 X_2) = \lambda_1 \phi(X_1) + \lambda_2 \phi(X_2)$; and normalized: $\phi(I) = 1$) functional on $\mathcal{A}$ one considers the two sided ideal J_ϕ in $\mathcal{A}$ defined by

$$J_\phi \equiv \{X \mid \phi(X^*X) = 0\}$$

Factorizing $\mathcal{A}$ by J_ϕ one introduces a scalar product on the quotient space $\mathcal{A}_{J_\phi}$ by

$$\langle \psi_X, \psi_Y \rangle \equiv \phi(X^*Y)$$

where ψ_X, ψ_Y denotes the equivalence classes determined by X, Y, and the representation $\pi_\phi: \mathcal{A} \to \mathcal{B}(\mathcal{H}_\phi)$ is defined on the completion of $\mathcal{A}_{J_\phi}$ by putting

$$\pi_\phi(X)(\psi_Y) \equiv \psi_{XY}$$

The von Neumann algebras are special ∗ subalgebras of $\mathcal{B}(\mathcal{H})$ which are closed not only in the norm but in some weaker topologies as well. We now define these topologies.

Let $\eta \in \mathcal{H}$ and $\eta_n, \zeta_n \in \mathcal{H}$ be such that

$$\sum_n \|\eta_n\|^2 < \infty \text{ and } \sum_n \|\zeta_n\|^2 < \infty$$

hold. Then the functions p_η, $w_{\{\eta\}}$ and $w_{\{\eta_n\}}$ defined respectively by

(i) $Q \to p_\eta(Q) = \|Q\eta\|$
(ii) $Q \to w_{\{\eta\}}(Q) = |\langle \eta, Q\eta \rangle|$
(iii) $Q \to w_{\{\eta_n\}}(Q) = \sum_n |\langle \eta_n, Q\zeta_n \rangle|$

are seminorms on $\mathcal{B}(\mathcal{H})$, and the locally convex topologies defined in $\mathcal{B}(\mathcal{H})$ by the seminorms (i)-(iii) (the weakest topologies in which every seminorm in the groups (i)-(iii) are continuous) are called *strong, weak and ultraweak* (operator) topologies, respectively.

Definition 6.3 Let $\mathcal{M}$ be a ∗-closed subalgebra of $\mathcal{B}(\mathcal{H})$ containing the unit element I. $\mathcal{M}$ is a *von Neumann algebra* if it is closed in the strong operator topology.

In particular, the von Neumann algebras are C^*-algebras, since they are closed in the uniform topology: if the sequence of elements Q_n in the von Neumann algebra $\mathcal{M}$ converges to Q in the operator norm, then Q_n also converges in the strong operator topology (since the latter is weaker than the uniform), thus $Q \in \mathcal{M}$.

Let $\mathcal{M} \subseteq \mathcal{B}(\mathcal{H})$ be an arbitrary subset of bounded operators, and let $\mathcal{M}'$ denote the operators in $\mathcal{B}(\mathcal{H})$ that commute with every element of $\mathcal{M}$, i.e.
$$\mathcal{M}' \equiv \{Q \in \mathcal{B}(\mathcal{H}) \mid QR = RQ, \; R \in \mathcal{M}\}$$
$\mathcal{M}'$ is called the first and $\mathcal{M}'' \equiv (\mathcal{M}')'$ the second *commutant* of $\mathcal{M}$. Obviosly, $\mathcal{M}'$ is a $*$-algebra containing the unit I element, furthermore, $\mathcal{M}' = \mathcal{M}'''$ for an arbitrary non-empty set $\mathcal{M}$ of operators. The following theorem is a key result in the theory of von Neumann algebras.

Proposition 6.1 (Neumann's double commutant Theorem) *Let $\mathcal{M}$ be a $*$-closed subalgebra of $\mathcal{B}(\mathcal{H})$ that contains I. Then the following are equivalent*

(i) $\mathcal{M} = \mathcal{M}''$

(ii) *$\mathcal{M}$ is closed in the strong, weak and ultraweak topologies in $\mathcal{B}(\mathcal{H})$.*

Proof: We prove the proposition for the strong operator topology only. Since $\mathcal{M} \subseteq \mathcal{M}''$ always, and since one can see easily that the commutant of a set of operators is always closed in the strong topology, it suffices to see that $\mathcal{M}$ is strongly dense in $\mathcal{M}''$. Let $Y \in \mathcal{M}''$. What we must show is that for any finite set $\xi_i \in \mathcal{H}$ of vectors there is an element $X \in \mathcal{M}$ such that $X\xi_i$ is arbitrary close to $Y\xi_i$ for all i. Let us suppose first that we have one vector ξ only. Let $\mathcal{K}$ be the closure of the vector space
$$\{X\xi \mid X \in \mathcal{M}\} \tag{6.1}$$
and let E be the projection onto this closed subspace. Then $X\mathcal{K} \subseteq \mathcal{K}$ for all $X \in \mathcal{M}$, which means that $XE = EXE$ and $X^*E = EX^*E$ holds and (by taking the adjoint of the last equality) it follows that $XE = EX$ for all $X \in \mathcal{M}$. Hence $E \in \mathcal{M}'$ and since $Y \in \mathcal{M}''$ we have $EY = YE$, which implies $Y\mathcal{K} \subseteq \mathcal{K}$. Since the identity I is in $\mathcal{M}$, the vector ξ is in $\mathcal{K}$, thus $Y\mathcal{K} \subseteq \mathcal{K}$ implies that $Y\xi \in \mathcal{K}$, i.e. $Y\xi$ is in the closure of (6.1), and so there is an $X \in \mathcal{M}$ such that $X\xi$ is close to $Y\xi$. The case of a finite set ξ_i can be reduced to the case of a single ξ by the following procedure. Consider $\overline{\mathcal{H}} \equiv \oplus_{i=1}^{i=n} \mathcal{H}$ on which $\mathcal{M}$ and Y act diagonally:
$$\overline{X}(\xi_1, \xi_2 \ldots \xi_n) = (X\xi_1, X\xi_2, \ldots X\xi_n)$$
$\overline{\mathcal{M}} = \{\overline{X} \mid X \in \mathcal{M}\}$ becomes then a $*$- algebra on $\overline{\mathcal{H}}$ and one can check that $\overline{\mathcal{M}}''$ is just the algebra of matrices that have a single operator fom $\mathcal{M}''$ at all diagonal entries. One can then apply the first part of the reasoning to this "amplification" $\overline{\mathcal{M}}, \overline{Y} \in \overline{\mathcal{M}}''$ and $(\xi_1, \xi_2 \ldots \xi_n) \in \overline{\mathcal{H}}$. □

As a corollary of the double commutant theorem we have the following characterization of von Neumann algebras:

Proposition 6.2 *Let $\mathcal{M}$ be a $*$-closed subalgebra of $\mathcal{B}(\mathcal{H})$ that contains I. Then $\mathcal{M}$ is a von Neumann algebra if and only if $\mathcal{M} = \mathcal{M}''$.*

The above proposition shows that the significance of the double commutant theorem is that it characterizes the von Neumann algebras in terms of the algebraic operations only, and so to prove that a selfadjoint set of operators is a von Neumann algebra, it suffices to show that it is closed with respect to the commutant operation. Also, the double commutant theorem tells us that a von Neumann algebra is closed in the weak and ultraweak topologies as well. Immediate consequences of the double commutant theorem are the following

1. The operator Q is in the von Neumann algebra $\mathcal{M}$ if and only if Q commutes with every selfadjoint element in $\mathcal{M}'$.
2. If $\mathcal{M}$ is a non-empty $*$-closed set of operators, then $\mathcal{M}'$ is a Neumann algebra.
3. If Q is a selfadjoint element in the von Neumann algebra $\mathcal{M}$, then every spectral projection of Q belongs to $\mathcal{M}$.

Definition 6.4 *The von Neumann algebra $\mathcal{M}$ is called a factor, if its center $\mathcal{Z}_\mathcal{M} \equiv \mathcal{M} \cap \mathcal{M}'$ is trivial, i.e. it consists of the elements λI only.*

The physical motivation behind considering operator algebras other than the algebra of all bounded operators is the following. In the Hilbert space formalism of quantum mechanics the bounded observables are represented by the selfadjoint part $\mathcal{B}(\mathcal{H})_{sa}$ of the set $\mathcal{B}(\mathcal{H})$ of all bounded linear operators on the Hilbert space $\mathcal{H}$. $\mathcal{B}(\mathcal{H})_{sa}$ inherits the linear structure in $\mathcal{B}(\mathcal{H})$ in a natural manner: $\mathcal{B}(\mathcal{H})_{sa}$ is a linear space (over the reals); however, the product (composition) of two observables is not, in general, an observable: unless A and B commute $(A, B, \in \mathcal{B}(\mathcal{H})_{sa})$ their product is not a selfadjoint operator. It was realized by Jordan, Wigner and von Neumann [79], [80] that the symmetric (or Jordan as it is now called) product defined by

$$A \bullet B = 1/2((A+B)^2 - A^2 - B^2) = 1/2(AB + BA) \qquad (6.2)$$

is selfadjoint even if A and B are non-commuting (selfadjoint) operators. The Jordan product is commutative, but non-associative in general. It has the following properties:

(1) $A \bullet B = B \bullet A$
(2) $A \bullet (B \bullet A^2) = (A \bullet B) \bullet A^2$
(3) $A \bullet (pB + rC) = pA \bullet B + qA \bullet C$ for all real numbers p, q
(I) $\| A \bullet B \| \leq \| A \| \| B \|$
(II) $\| A^2 \| = \| A \|^2$
(III) $\| A^2 \| \leq \| A^2 + B^2 \|$

A real linear space in which a product • is defined having the properties (1)-(3) is called a *real Jordan algebra*. A Jordan algebra equipped with a norm such that the algebra is complete with respect to that norm and such that the product has the property (I) above is called a *Jordan Banach algebra*; finally, a Jordan Banach algebra satisfying also (II)-(III) is called a *JB algebra*.

Thus the (bounded) observables on a Hilbert space form a JB algebra. The main idea of the so called "algebraic approach" to quantum mechanics is that in modeling the quantum system it is the JB structure of the observables that is essential, therefore, this should be taken as a primitive concept, and the properties (1)-(3) and (I)-(III) should be postulated. Clearly, the selfadjoint part of a C^*-algebra $\mathcal{A}$ is a JB algebra with the product defined in complete analogy with (6.2). Thus, all C^*-algebras, and in particular all von Neumann algebras define a Jordan algebra. It will be seen in the next section that there exist von Neumann algebras that have a highly non-$\mathcal{B}(\mathcal{H})$-type-character, hence their Jordan algebra structure, too, is far from the usual structure of $\mathcal{B}(\mathcal{H})$. It must be emphasized that the existence of these non-$\mathcal{B}(\mathcal{H})$-type algebras is not only a mathematical delicacy: there exist very real physical situations in which these structures appear as models of the observable quantities (see the remarks after the table summarizing the classification of factors, next section, and the Section 10.1).

6.2. Von Neumann lattices

Proposition 6.3 *The lattice of projections of a von Neumann algebra is a complete, orthomodular lattice with respect to the lattice operations inherited from $\mathcal{P}(\mathcal{H})$; furthermore, $\mathcal{P}(\mathcal{M})$ generates $\mathcal{M}$ in the sense that $\mathcal{P}(\mathcal{M})'' = \mathcal{M}$.*

Proof: $\mathcal{P}(\mathcal{M})$ contains 0 and I and it is $*$-closed, thus $\mathcal{P}(\mathcal{M})'$ is a von Neumann algebra. $\mathcal{P}(\mathcal{M})''$ also is a von Neumann algebra, and because $\mathcal{P}(\mathcal{M}) \subseteq \mathcal{M}$ we have $\mathcal{P}(\mathcal{M})' \supseteq \mathcal{M}'$ (the commutant of a smaller set is larger). If Q is in the von Neumann algebra $\mathcal{P}(\mathcal{M})'$, then all spectral projections of Q are in $\mathcal{P}(\mathcal{M})'$, too, hence Q commutes with the spectral projections of every selfadjoint element R in $\mathcal{M}$ since these lie in $\mathcal{P}(\mathcal{M})$. Consequently, Q commutes with every selfadjoint element, and hence with every element in $\mathcal{M}$. This means that $\mathcal{P}(\mathcal{M})' \subseteq \mathcal{M}'$. It follows that $\mathcal{P}(\mathcal{M})' = \mathcal{M}'$, and so $\mathcal{P}(\mathcal{M})'' = \mathcal{M}'' = \mathcal{M}$.

We now show that $\mathcal{P}(\mathcal{M})$ is a complete lattice. Let $\mathcal{S} \subseteq \mathcal{P}(\mathcal{M})$ be an arbitrary set of projections and P be the least upper bound (joint) of the projections in $\mathcal{S}$. The orthocomplementation $\perp$ is defined in $\mathcal{P}(\mathcal{H})$ algebraically ($A^\perp = I - A$), and $\wedge$ can be defined in terms of $\vee$ and $\perp$ by the De Morgan rule:

$$\wedge \mathcal{S} = (\vee \{A^\perp \mid A \in \mathcal{S}\})^\perp$$

and this implies that it is enough to show that $P \in \mathcal{P}(\mathcal{M})$. Choose an $X \in \mathcal{M}'$. If $E \in \mathcal{S}$ then $XE = EX$. Consider $PX - XP$ and multiply it on the right by E. Using $PE = E$ (since $E \leq P$) one obtains

$$(PX - XP)E = PXE - XPE = PEX - XPE = EX - XE = 0$$

that is the subspace E projects to is contained in the null space of $PX - XP$. Let $ker(PX - XP)$ be the projection onto this null space. Then

$$ker(PX - XP) \geq E$$

and since this holds for any $E \in \mathcal{S}$ and because P is the least upper bound of $\mathcal{S}$, we have
$$P \leq ker(PX - XP)$$

It follows that
$$(PX - XP)P = 0$$

i.e. $PXP = XP$. Since X was arbitrary, we can take X^*, and repeating the same steps as for X we arrive at $PX^*P = X^*P$, which implies (after taking the adjoint) $PXP = PX$, thus $XP = PX$, and so P commutes with every element in $\mathcal{M}'$, i.e. $P \in \mathcal{M}'' = \mathcal{M}$. □

This Proposition 6.3 is of central importance in the theory of von Neumann algebras. The fact that the projection lattice generates the algebra indicates that by investigating this lattice structure one gets insight into the structure of the algebra itself. Indeed, von Neumann and Murray used this lattice structure to classify von Neumann algebras. In what follows this classification is described briefly. To facilitate the overview the proofs of the most important facts in connection with the classification theory are postponed and are collected in the Appendix to this chapter.

The key concept in the classification is the equivalence of projections: two projections A and B in $\mathcal{M}$ are called equivalent (notation: $A \sim B$) with respect to the algebra $\mathcal{M}$ if there is an operator ("partial isometry") in $\mathcal{M}$ that takes the vectors in $A^\perp$ into zero and is an isometry between the subspaces A and B. This equivalence can be expressed by saying that the dimension (with respect to $\mathcal{M}$) of the subspace A projects to is equal to the dimension (with respect to $\mathcal{M}$) of the subspace B projects to. The relation $\sim$ is an equivalence relation in $\mathcal{P}(\mathcal{M})$ (Proposition 6.10). Let $\mathcal{P}(\mathcal{M})_\sim$ be the set of equivalence classes. With the help of $\sim$ one can introduce a $\preceq$ partial ordering in $\mathcal{P}(\mathcal{M})$ ("Cantor-Bernstein type theorem" Proposition

6.11): $A \preceq B$ if there is a projection B' that is smaller (with respect to $\leq$) than B and which is equivalent with A, that is: $A \sim B' \leq B$. Intuitively $A \preceq B$ means: the (relative) dimension of A is not greater than the (relative) dimension of B. From the point of view of classification of von Neumann algebras the important fact concerning $\preceq$ is that given any two projections A, B, they can be cut *by a projection commuting with every element in* $\mathcal{M}$ into two pieces that can be compared in the ordering $\preceq$. That is ("Comparison Theorem", Proposition 6.12): for any two $A, B \in \mathcal{P}(\mathcal{M})$ there is a projection $Z \in \mathcal{P}(\mathcal{M}) \cap \mathcal{P}(\mathcal{M})''$ such that

$$ZAZ \preceq ZBZ \text{ and } (I-Z)B(I-Z) \preceq (I-Z)A(I-Z)$$

It follows that if $\mathcal{M}$ is a factor von Neumann algebra, then $\mathcal{P}(\mathcal{M})_\sim$ is *totally* ordered with respect to $\preceq$: either $A \preceq B$, or $B \preceq A$ holds for any A, B. Since $\preceq$ is defined algebraically, it follows that the ordering is algebra invariant: two factor von Neumann algebras can not be isomorphic if the orderings of the corresponding $\mathcal{P}(\mathcal{M})_\sim$ is different.

To determine the order type of $\mathcal{P}(\mathcal{M})_\sim$ a key concept is the finiteness of projections: A projection is *finite*, if it is not equivalent to any proper subprojection of itself. That is, A is finite if from $A \sim B \leq A$ it follows that $A = B$.

Proposition 6.4 *If $\mathcal{M}$ is a factor von Neumann algebra, then there exists a map d (unique up to multiplication by a constant) defined on $\mathcal{P}(\mathcal{M})$ and taking its values in the closed interval $[0, \infty]$ and which has the following properties:*

(i) $d(A) = 0$ *if and only if* $A = 0$
(ii) *If* $A \perp B$, *then* $d(A + B) = d(A) + d(B)$
(iii) $d(A) \leq d(B)$ *if and only if* $A \preceq B$
(iv) $d(A) < \infty$ *if and only if* A *is a finite projection*
(v) $d(A) = d(B)$ *if and only if* $A \sim B$
(vi) $d(A) + d(B) = d(A \wedge B) + d(A \vee B)$

The map d is called the *dimension* function on $\mathcal{P}(\mathcal{M})$. This proposition implies that the order type of $\mathcal{P}(\mathcal{M})_\sim$ can be read off the order type of the range of the function d. Murray and von Neumann determined the possible ranges of d in [104]. The result is shown in the table below (by chosing suitable normalization of the function d).

range of d	type of factor $\mathcal{N}$	example	the lattice $\mathcal{P}(\mathcal{N})$
$\{0, 1, 2, \ldots n\}$	I_n	$\mathcal{B}(\mathcal{H}_n)$ dim $\mathcal{H}_n = n$	modular, atomic non-distributive if $n \geq 2$
$\{0, 1, 2, \ldots \infty\}$	I_∞	$\mathcal{B}(\mathcal{H})$ dim $\mathcal{H} = \infty$	orthomodular, non-modular atomic
$[0, 1]$	II_1	$\otimes_n M_2$	modular, non-atomic
$\{x \mid 0 \leq x \leq \infty\}$	II_∞		non-modular, non-atomic
$\{0, \infty\}$	III	$\mathcal{A}(W)$	non-modular non-atomic

This table shows that besides the well known von Neumann algebra $\mathcal{B}(\mathcal{H})$ of all bounded operators there are other types of von Neumann algebras. These other types have properties *radically* different from the properties of $\mathcal{B}(\mathcal{H})$. For instance, in sharp contrast to the type **I** case, the projection lattice of a type II_1 algebra has no atoms and it is not only orthomodular but also modular, which $\mathcal{P}(\mathcal{H})$ is not (Proposition 4.4). The modularity of the lattice $\mathcal{P}(\mathcal{M})$ of a type II_1 factor $\mathcal{M}$ follows from the Proposition 3.3 and from the existence of a finite d; however, a direct proof of modularity of the lattice of a finite (not necessarily factor) von Neumann algebra also will be given in the Appendix to this chapter (see Proposition 6.14).

That there are no atoms in the lattice $\mathcal{P}(\mathcal{M})$ of a type II_1 factor also is an easy consequence of the existence of the dimension function d: Take an element $0 \neq A \in \mathcal{P}(\mathcal{M}))$. Then $d(A) > 0$ and chose a projection $B \in \mathcal{P}(\mathcal{M}))$ such that $d(B) = \frac{1}{2}d(A)$. Then $B \preceq A$, thus there exists an A' projection such that $A' \leq A$ and $A' \sim B$, hence $d(A') = d(B) = \frac{1}{2}$. The projection A' is thus a non-zero projection strictly smaller than A, consequently A is not an atom. There are no smallest (non-zero) dimensional projections in the lattice of a type II_1 von Neumann algebra, although every projection has a unique (finite) dimension. This justifies calling the theory of II_1 factors the theory of continuous dimension.

The type **III** algebras are even farther away from $\mathcal{B}(\mathcal{H})$ than are the type II_1 ones: they contain no finite projection at all. It is worth mentioning that at the time of the discovery (the year 1936) of the classification of factors it was not even known whether type III factors exist. It was only four years later that von Neumann was able to construct the first example [172], and it was only in the mid sixties that the existence of a continuum number of mutually non-isomorphic factors was proven [114]. Physical examples of type III factors are the algebras $\mathcal{A}(W)$ of observables that are localized (in the sense of relativistic quantum field theory) in wedge regions (Proposition 10.2). Type II_1 algebras had been constructed by Birkhoff and von Neumann in 1936, however. An example of a type II_1 factor is given at the end of this chapter.

A distinguishing feature of a type II_1 factor $\mathcal{M}$ is that the dimension function d defined on $\mathcal{P}(\mathcal{M})$ can be extended to a *unique, finite* trace on $\mathcal{M}$, just like the dimension function on the projection lattice of a finite dimensional Hilbert space.

Definition 6.5 The map $\tau : \mathcal{M}^+ \to [0,\infty]$ on a von Neumann algebra $\mathcal{M}$ is called *trace* if it has these properties:
(i) $\tau(X+Y) = \tau(X) + \tau(Y) \qquad X, Y \in \mathcal{M}^+$
(ii) $\tau(\lambda X) = \lambda \tau(X) \qquad X \in \mathcal{M}^+, \lambda \geq 0$
(iii) $\tau(X^*X) = \tau(XX^*) \qquad X \in \mathcal{M}$

The trace τ is called *faithful* if $\tau(X) > 0$ for all $0 \neq X \geq 0$. The trace τ is called *finite* if $\tau(X) < \infty$ for all $X \in \mathcal{M}^+$.

For instance the usual trace

$$\tau(X) = Tr(X) = \sum_i \langle \eta_i, X\eta_i \rangle \tag{6.3}$$

(η_i being an orthonormed basis in $\mathcal{H}$) defines a trace on $\mathcal{B}(\mathcal{H})$ in the sense of the above definition of trace, which is, however, not finite unless $\mathcal{H}$ is finite dimensional. If $\mathcal{H}$ is finite dimensional, then the summation in (6.3) is finite, and $\tau(X)$ is then the "Spur" of the matrix X.

Remark: Since $\mathcal{M}^+$ spans the whole algebra $\mathcal{M}$ linearly, a finite trace can always be extended from $\mathcal{M}^+$ to $\mathcal{M}$, hence a finite trace is just a positive functional possessing property (iii) of the trace. This property implies that a trace has this feature:

$$\tau(AB) = \tau(BA)$$

for any elements $A, B \in \mathcal{M}$. To show this assume first that A, B are selfadjoint and set $C = A + iB$, $C^* = A - iB$. Then

$$\tau(C^*C) = \tau(AA + iAB - iBA + BB) \tag{6.4}$$
$$\tau(CC^*) = \tau(AA - iAB + iBA + BB) \tag{6.5}$$
$$\tag{6.6}$$

which implies

$$0 = \tau(C^*C) - \tau(CC^*) = 2i\tau(AB - BA)$$

Since any $Q \in \mathcal{M}$ can be written as the linear sum of selfadjoint elements:

$$Q = Q_1 + iQ_2 \text{ with } Q_1 = \frac{1}{2}(Q + Q^*) \text{ and } Q_2 = i\frac{1}{2}(Q - Q^*)$$

the general case follows from the $\tau(AB) = \tau(BA)$ with selfadjoint A, B. In particular it holds that τ is unitary invariant:

$$\tau(X) = \tau(XUU^*) = \tau(U^*XU) \quad (X \in \mathcal{M})$$

Definition 6.6 A family $S = \{\tau_i\}_{i \in J}$ of traces on a von Neumann algebra is called *sufficient* if for every $0 \neq X \in \mathcal{M}^+$ there is an $i \in J$ such that $\tau_i(X) \neq 0$.

A natural extension of the concept of a numerical valued trace is the following concept of trace:

Definition 6.7 The map $T: \mathcal{M} \to \mathcal{Z}_\mathcal{M}$ is called *center valued trace* if
(i) $T(X + Y) = T(X) + T(Y) \quad T(\lambda X) = \lambda T(X)$
(ii) $T(X^*X) = T(XX^*)$
(iii) $T(I) = I$
(iv) $T(ZX) = ZT(X) \quad X \in \mathcal{M}, Z \in \mathcal{Z}_\mathcal{M}$
(v) $T(X^*X) > 0$ if $X \neq 0$

The existence of a center valued trace on a von Neumann algebra is characteristic of *finite* von Neumann algebras.

Definition 6.8 A von Neumann algebra is finite if its identity, thus every projection in it is finite.

Proposition 6.5 *Equivalent are the following:*

(i) *The von Neumann algebra $\mathcal{M}$ is finite;*
(ii) *There exists a sufficient family of traces on $\mathcal{M}$;*
(iii) *There exist a center valued trace on $\mathcal{M}$.*

One shows that the dimension function on a type II_1 factor can be extended to a trace by proving first that there exists a finite (normalized) trace on a type II_1 factor, (the sketch of this proof is given in the next section). It is then easy to see that the trace has the characteristic property (v) in Proposition 6.4 of the dimension function (see Proposition 6.13).

An example of a type II_1 *factor*

Let $\mathcal{N}_n \equiv \otimes_{i=1}^n M_k$ be the n-fold tensor product of the matrix algebra of k-by-k two complex matrices by itself, and let τ_n be the normalized trace on $\mathcal{N}_n$, which is the unique tracial state on $\mathcal{N}_n$. This state is defined by

$$\tau_n(X_1 \otimes X_2 \otimes \ldots X_n) \equiv \tau_k(X_1)\tau_k(X_2)\ldots\tau_k(X_n) \qquad X_i \in M_k$$

where $\tau_k = \frac{Tr}{k}$ is the normalized trace on M_k. The algebra $\mathcal{N}_n$ can be viewed as a subalgebra of $\mathcal{N}_{n+1}$ by the identification $X \leftrightarrow X \otimes I$, and the union $\mathcal{A} = \cup_n^\infty \mathcal{N}_n$ is an infinite dimensional, normed $*$-algebra. The state τ defined by

$$\tau(X_n) = \tau_n(X_n) \qquad (X_n \in \mathcal{N}_n)$$

is a tracial state on $\mathcal{A}$. We can define an inner pruduct in $\mathcal{A}$ by

$$\langle X, Y \rangle = \tau(X^*Y) \qquad X, Y \in \mathcal{A}$$

and $\mathcal{A}$ becomes a pre-Hilbert space, completion of which is denoted by $\mathcal{H}_\tau$. Let Φ denote the vector in $\mathcal{H}_\tau$ corresponding to the unit in $\mathcal{A}$. One then can define a representation of $\mathcal{A}$ on $\mathcal{H}_\tau$ by

$$\pi_\tau(X)Y = XY \qquad (X, Y \in \mathcal{A})$$

The operator $\pi_\tau(X)$ is densely defined and can be extended by continuity to a bounded operator on $\mathcal{H}_\tau$ (also denoted by $\pi_\tau(X)$). Also,

$$\langle \Phi, \pi_\tau(X)\Phi \rangle = \tau(X) \qquad (X \in \mathcal{A})$$

holds. We claim that the von Neumann algebra $\mathcal{N} = \pi_\tau(\mathcal{A})''$ is a type II_1 factor.

We show first that it is a factor. What one has to show is that if $XY = YX$ for every $Y \in \mathcal{N}$, then $X = \lambda I$. The proof of this makes use of the existence of a conditional expectation T_n from $\mathcal{N}$ onto $\pi_\tau(\mathcal{N}_n)$ for every n and the fact that $\mathcal{N}_n$ is a factor. T_n is defined as follows (see Proposition 8.6 for details). $\mathcal{N}_n$ is a closed linear subspace in $\mathcal{H}_\tau$, so there is an orthogonal projection P_n to this subspace. Let $T_n: \mathcal{N} \to \pi_\tau(\mathcal{N}_n)$ be the conditional expectation defined by

$$T_n(X)Y_n = P_n X Y_n \qquad (Y_n \in \mathcal{N}_n)$$

If $XY = YX$ (for every $Y \in \mathcal{N}$), then also $X\pi_\tau(Y) = \pi_\tau(Y)X$ for every $Y \in \mathcal{A}$. If $Y \in \mathcal{N}_m$ and $n > m$, then

$$T_n(X\pi_\tau(Y)) = T_n(\pi_\tau(Y)X)$$

However, by the properties of the conditional expectation (see Proposition 8.6) we have

$$T_n(X\pi_\tau(Y)) = T_n(X)\pi_\tau(Y)$$
$$T_n(\pi_\tau(Y)X) = \pi_\tau(Y)T_n(X)$$

Hence it holds that

$$T_n(X)\pi_\tau(Y) = \pi_\tau(Y)T_n(X)$$

Since $\mathcal{N}_m$ is a factor, it follows that $T_n(X) = \lambda I_n$. Using the continuity property of the conditional expectations $T_n(X) \to X$ ($n \to \infty$) we obtain $X = \lambda I$.

To see that $\mathcal{N}$ is type II_1 we must show the existence of a faithful, normal trace on $\mathcal{N}$. The state τ is normal since it is given by a vector and it also has the tracial property

$$\tau(XY - YX) = 0 \quad (X, Y \in \mathcal{N})$$

because $\tau(XY - YX) = 0$ holds for $X, Y \in \pi_\tau(\mathcal{A})$. So one only has to show that τ is faithful, i.e. $\tau(X) = 0$ and $A \geq 0$ implies $A = 0$. $\tau(X) = 0$ is equivalent to $X^{1/2}\Phi = 0$, and by the Kaplansky's density theorem there exists a series $X_n \in \mathcal{N}_n$ such that

$$\| X_n \| \leq \| X^{1/2} \|$$

and $\pi_\tau(X_n)$ converges to $X^{1/2}$ strongly. For any $Y \in \mathcal{A}$ one has then

$$\| X^{1/2} Y \Phi \|^2 = \lim_n \| \pi_\tau(X_n) Y \Phi \|^2 \tag{6.7}$$
$$= \lim_n \tau(Y^* X_n^* X_n Y) \leq \lim_n \| Y \|^2 \tau(X_n^* X_n) \tag{6.8}$$
$$= \| Y \|^2 \tau(Y) = 0 \tag{6.9}$$

This shows that $X^{1/2} = 0$ on a dense set, consequently $X = 0$.

So $\mathcal{N}$ is a type II_1 factor. By construction $\mathcal{N}$ contains a series of finite dimensional matrix algebras whose union is dense in $\mathcal{N}$. Such algebras are called *approximately finite dimensional* (AFD) or *hyperfinite*. From the above proof it is also clear that one could start with any matrix algebra M_k ($k = 2, 3 \ldots$) and construct a type II_1 factor, the value of the number k is irrelevant. In fact more is true: the above construction gives isomorphic type II_1 factors for any k, and it can be proved that (up to isomorphism) there is only one hyperfinite II_1 factor, it is denoted by $\mathcal{R}$. Note also that the range of the dimension function τ on $\mathcal{N}_n$ is

$$\{\frac{m}{k^n} \mid 0 \leq m \leq k^n\}$$

which is $[0,1]$ in the limit $k \to \infty$.

Summing up: A selfadjoint set of operators on a Hilbert space is a von Neumann algebra if it is equal to its double commutant. A von Neumann algebra is generated by its projections, the set of which is an orthomodular lattice. The factor von Neumann algebras can be classified on the basis of the range of the dimension function defined on (the equivalence classes) of projections. A not necessarily factor von Neumann algebra is called finite if it does not contain infinite projections. A von Neumann algebra is finite if and only if it has a faithful center valued trace. The center valued trace is unique if the algebra is a finite factor, and the restriction of this unique trace to the projection lattice coincides with the unique numerical dimension function in this case.

It is natural now to ask the question whether a von Neumann lattice $\mathcal{P}(\mathcal{M})$ can be interpreted as the logic of the quantum system whose observables are described by the selfadjoint elements of the algebra $\mathcal{M}$. The answer is yes: One can form the set $\mathcal{F}^{\mathcal{M}}$ of elementary sentences just like in the case of of a Hilbert lattice (Chapter 5), with the obvious modification that Q in the elementary sentence $sent(Q,d,1)$ is now assumed to be a selfadjoint element in $\mathcal{M}$. Since the spectral projections $\mathrm{P}^Q(d)$ of Q lie in $\mathcal{M}$, the vectors ξ in $\mathrm{P}^Q(d) \in \mathcal{P}(\mathcal{M})$ again make the elementary sentence $sent(Q,d,1)$ true, and all the definitions and steps in Section 5.3 that prove that $\mathcal{F}^q_{\leftrightarrow}$ is isomorphic to the lattice $\mathcal{P}(\mathcal{H})$ can be repeated to show that $\mathcal{F}^{\mathcal{M}}_{\leftrightarrow}$ is isomorphic to the von Neumann lattice $\mathcal{P}(\mathcal{M})$. Note that the essential property of a von Neumann algebra used in this reasoning is that all the spectral projections of a selfadjoint element in a von Neumann algebra belong to the von Neumann algebra. If this were not the case, clearly then the projections could not be viewed as collections of interpretations making the typical elementary sentences true.

6.3. Appendix: proofs of propositions related to the classification theory of von Neumann algebras

Let Q be any element in a von Neumann algebra $\mathcal{M}$. The smallest projection A in $\mathcal{M}$ such that $AQ = Q$ is called the *left support* of Q and is denoted by $s_l(Q)$; the smallest projection B in $\mathcal{M}$ such that $QB = Q$ is called the *right support* of Q and is denoted by $s_r(Q)$. If $\mathcal{M}$ acts on the Hilbert space $\mathcal{H}$, then $s_l(Q)$ is the projection onto the closure of the range of Q:

$$s_l(Q) = closure \ \{Q\xi \mid \xi \in \mathcal{H}\}$$

and $s_r(Q)$ is the projection onto the closure of the range of Q^*. If $Q = A$ is a projection, then $s_l(A) = A = s_r(A)$. The polar decomposition theorem tells us that if Q is any element in a von Neumann algebra $\mathcal{M}$, then

$$Q = U(H^*H)^{1/2}$$

where U is a partial isometry with initial projection $s_l(H^*)$ and final projection $s_l(H)$. This means that

Proposition 6.6 *For any Q in a von Neumann algebra it holds that*

$$s_l(Q) \sim s_r(Q) = s_l(Q^*)$$

An immediate consequence of the above proposition is the

Proposition 6.7 (Paralellogram rule, Kaplansky formula) *Let A and B be arbitrary projections in the von Neuman algebra $\mathcal{M} \subseteq \mathcal{B}(\mathcal{H})$. Then*

$$(A \vee B) - B \sim A - (A \wedge B) \quad (6.10)$$

Proof: Using the little Lemma below we show that $A \vee B - B$ is the range projection (left support) of $(I - B)A$ and $A - A \wedge B$ is the range projection of $A(I - B)\big(= [(I - B)A]^*\big)$, and then the equivalence is implied by Proposition (6.6).

Lemma: If E and F are projections on the Hilbert space $\mathcal{H}$, then

$$range(E + F) = E \vee F \qquad range(EF) = E - E \wedge (I - F) \quad (6.11)$$

By (6.11) we have

$$range(A(I - B)) = A - A \wedge (I - I - B) = A - A \wedge B$$

$$\begin{aligned}
range((I-B)A) &= I - B - (I - B) \wedge (I - A) \\
= [(I - (I - B) \wedge (I - A))^\perp]^\perp - B &= ((I - B) \wedge (I - A))^\perp - B \\
= (I - B)^\perp \vee (I - A)^\perp - b &= B \vee A - B = A \vee B - B
\end{aligned}$$

Proof of Lemma: Since

$$\begin{aligned}
\| E\xi \|^2 + \| F\xi \|^2 &= \langle E\xi, \xi \rangle + \langle F\xi, \xi \rangle \\
&= \langle (E + F)\xi, \xi \rangle
\end{aligned}$$

for every $\xi \in \mathcal{H}$, it follows that $(E + F)\xi = 0$ if and only if $E\xi = F\xi = 0$, hence

$$ker(E + F) = ker(E) \wedge ker(F) = (I - E) \wedge (I - F) \quad (6.12)$$

using the fact that the range of an operator Q is equal to the orthogonal complement of the kernel of its adjoint:

$$range(Q) = I - ker(Q^*) \tag{6.13}$$

(6.12) implies

$$range(E+F) = I - ker(E+F) = I - (I-E) \wedge (I-F)$$
$$= E \vee F$$

Because of (6.13), to show the second equality in (6.11) it is enough to see that

$$ker(FE) = I - E + E \wedge (I-F)$$

If $\xi \in ker(FE)$, then

$$E\xi = E\xi - FE\xi = (I-F)E\xi \in (E \wedge (I-F))$$

therefore

$$\xi = (I-E)\xi + E\xi \in (I - E + E \wedge (I-F))$$

Conversely, if $\xi \in (I-E+E\wedge(I-F))$, then ξ can be written as $\xi = \eta + \theta$ such that

$$\eta = (I-E)\eta \text{ and } \theta = E\theta = (I-F)\theta$$

and so

$$FE\xi = FE\eta + FE\theta = FE(I-E)\eta + F(I-F)\theta = 0$$

Let A be a projection in a von Neumann algebra $\mathcal{M}$ and let $\{P_i\}$ be a family of *central* projections in $\mathcal{M}$ (i.e. $P_i \in \mathcal{Z}_\mathcal{M}$) such that $P_i A = 0$. Then $PA = 0$ where $P \equiv \vee P_i$. The projection $P^\perp = I - P$ is called the *central support (central carrier)* of A and it is denoted by $z(A)$. Equivalently, P is the intersection (greatest lower bound) of all central projections E such that $EA = A$.

Proposition 6.8 *Two projections A, B in a von Neumann algebra $\mathcal{M}$ have equivalent non-zero subprojections if and only if $z(A)z(B) \neq 0$.*

Proof: Assume that $z(A)z(B) = 0$, $A_0 \leq A, B_0 \leq B$ and $A_0 \sim B_0$. Then there is a partial isometry V such that $V^*V = A_0$ and $VV^* = B_0$. Since V maps A_0 onto B_0 and $B_0 = VV^*$ is a projection onto B_0, it holds that $V = V^*V$, and since $B_0 \leq B \leq z(B)$ and $A_0 \leq A \leq z(A)$ we have

$$V = VV^*V = VV^*VV^*V =$$
$$B_0 V A_0 = B_0 z(B) V z(A) A_0 = B_0 V A_0 z(B) z(A) = 0$$

Thus $0 = A_0 = B_0$. Conversely, assume that $z(B)z(A) \neq 0$. We show first that $A\mathcal{M}B \neq \{0\}$ in this case. Indeed, assume that $A\mathcal{M}B = \{0\}$ and consider
$$\mathcal{J} \equiv \{x \in \mathcal{M} \mid A\mathcal{M}x = \{0\}\}$$
Then $\mathcal{J}$ is an ultraweakly closed ideal in $\mathcal{M}$, and since any such ideal has the form $\mathcal{J} = \mathcal{M}Z$ where Z is a central projection, there exist such a Z and, B belonging to $\mathcal{J}$ by assumption, we have $B \leq Z$. It follows that $z(B) \leq Z$. By assumption we also have $AIZ = 0 = AZ$, hence $z(B)A = 0$, which implies $A \leq I - z(B)$. Therefore $z(A) \leq I - z(B)$ i.e. $z(A)$ and $z(B)$ are orthogonal, contrary to $z(B)z(A) \neq 0$. So $A\mathcal{M}B \neq \{0\}$. Let now Q be a nonzero element in $A\mathcal{M}B \neq \{0\}$. Then $AQ'B = Q$ for some $Q' \in \mathcal{M}$ and
$$Q = AQ'B = AAQ'BB = AQB$$
i.e. $Q = AQB$ and we have $0 \neq s_l(Q) \leq A$ and $0 \neq s_r(Q) \leq B$, furthermore $s_l(Q) \sim s_r(B)$ by Proposition (6.6). $\square$

As a corollary to the above proposition we have

Proposition 6.9 *Each pair of non-zero projections (A, B) in a factor von Neumann algebra have equivalent non-zero subprojections.*

Proof: If $A \neq 0 \neq B$, then $z(A) \neq 0 \neq z(B)$. Since the only nonzero projection in a factor is I, it follows that $z(A) = I = z(B)$, and the statement of the proposition follows from the preceding Proposition 6.8. $\square$

Proposition 6.10 *The relation $\sim$ is an equivalence relation on $\mathcal{P}(\mathcal{M})$.*

Proof: We must show that $\sim$ is (i) reflexive: $A \sim A$; (ii) transitive: if $A \sim B$ and $B \sim C$, then $A \sim C$; (iii) symmetric: if $A \sim B$, then $B \sim A$. Assume $A \sim B$ and $B \sim C$. Then there are partial isometries V and W such that
$$\begin{aligned} A &= V^*V \\ B &= VV^* \\ B &= W^*W \\ C &= WW^* \end{aligned}$$

It follows that
$$\begin{aligned} (V^*)^*V^* &= B \\ V^*(V^*)^* &= A \end{aligned}$$

thus $B \sim A$, i.e. (iii) holds. Since $A = AA^* = A^*A$ the projection A is a partial isometry with initial and final projection A, so $A \sim A$. To see (ii) consider

$$(WV)^*WV = V^*W^*WV = V^*BV = V^*VV^*V = AA = A$$
$$WV(WV)^* = WVV^*W^* = WBW^* = WW^*WW^* = CC = C$$

That is $A \sim C$. □

Proposition 6.11 (Cantor-Bernstein theorem for projections) *If A and B are projections in a Neumann lattice $\mathcal{P}(\mathcal{M})$, then $A \preceq B$ and $B \preceq A$ implies $A \sim B$.*

Proof: The idea of the proof is to decompose A and B into sums of mutually orthogonal, pairwise equivalent subprojections, and to use the following property of the $\sim$ equivalence relation:

"Complete divisibility" of $\sim$: If $\{A_i\}$ and $\{B_i\}$ ($i \in \mathcal{I}$) are orthogonal families of projections in a von Neumann algebra such that $A_i \preceq B_i$ for all i, then

$$A = \sum_i A_i \preceq \sum_i B_i = B$$

If $A_i \sim B_i$ for all i, then

$$\sum_i A_i \sim \sum_i B_i$$

To see the complete divisibility assume $A_i \sim B_i$ for all i, and let V_i be the partial isometry that implements the equivalence of A_i and B_i. Let us define V to be the linear extension of the map that is equal to V_i on the range of A_i and which is zero on the range of $I - \sum_i A_i$. (That is, $V \equiv \sum_i V_i$, where the sum is understood in the sense of strong topology.) Then V sets up the equivalence of $A = \sum_i A_i$ and $B = \sum_i B_i$. If $A_i \preceq B_i$, then for every i, $A_i \sim D_i \leq B_i$ for some D_i, and then $A \sim \sum_i D_i \leq B$, i.e. $A \preceq B$.

Let now V and W be partial isometries setting up the equivalence between A and a subprojection B_1 of B and B and a subprojection A_1 of A:

$$V^*V = A \qquad VV^* = B_1 \leq B \qquad (6.14)$$
$$W^*W = B \qquad WW^* = A_1 \leq A \qquad (6.15)$$

The partial isometry V maps the range of A isometrically onto the range of B_1, thus the subprojection A_1 of A is mapped by V isometrically onto a subprojection B_2 of B_1. This means algebraically that

$$(VA_1)^*VA_1 = A_1V^*VA_1 = A_1AA_1 = A_1 \qquad (6.16)$$
$$VA_1(VA_1)^* = VA_1A_1V^* = VA_1V^* = B_2 \qquad (6.17)$$

Similarly, the partial isometry W maps the range of B isometrically onto the range of the subprojection A_1, thus W maps B_1 onto a subprojection A_2 of A, and we have

$$(WB_1)^*WB_1 = B_1W^*WB_1 = B_1BB_1 = B_1 \quad (6.18)$$
$$WB_1(WB_1)^* = WB_1B_1W^* = WB_1W^* = A_2 \quad (6.19)$$

Furthermore, the projection $A - A_1$ is equivalent to the projection $B_1 - B_2$, since the partial isometry $V(A - A_1)$ sets up the equivalence between $A - A_1$ and $B_1 - B_2$:

$$\begin{aligned}
&[V(A - A_1)]^*V(A - A_1) \\
&= (A - A_1)V^*V(A - A_1) = (A - A_1)A(A - A_1) = A - A_1 \\
&V(A - A_1)[V(A - A_1)]^* \\
&= V(A - A_1)(A - A_1)V^* = V(A - A_1)V^* \\
&\quad = VAV^* - VA_1V^* = VAAV^* - VA_1A_1V^* \\
&\quad = VV^*VV^*VV^* - VA_1(VA_1)^* \\
&\quad = B_1B_1B_1 - B_2 = B_1 - B_2
\end{aligned}$$

Thus, starting with A_1 and B_1, existence of which is ensured by the assumption of the proposition, we have cut out two equivalent pieces $((A - A_1)$ and $(B_1 - B_2))$ from A and B, and by taking the image of B_1 under W we have an A_2, so the process can be repeated starting now with A_2, B_2 in place of A_1 and B_1. In this way we can construct two decreasing sequences of projections

$$\begin{aligned}
A &= A_0 \geq A_1 \geq A_2 \ldots A_n \ldots \\
B &= B_0 \geq B_1 \geq B_2 \ldots B_n \ldots
\end{aligned}$$

by the definitions

$$\begin{aligned}
A_{n+1} &\equiv WB_nW^* \quad n = 0, 1, \ldots \\
B_{n+1} &\equiv VA_nV^* \quad n = 0, 1, \ldots
\end{aligned}$$

Putting $A_\infty = \lim A_n = \wedge_n A_n$ and $B_\infty = \lim B_n = \wedge_n B_n$ we have the following relations

$$\begin{aligned}
V(A_n - A_{n+1})V^* &= B_{n+1} - B_{n+2} \\
W(B_n - B_{n+1})W^* &= A_{n+1} - A_{n+2} \\
VA_\infty V^* = B_\infty &\quad WB_\infty W^* = A_\infty
\end{aligned}$$

It follows that $A_n - A_{n+1}$ is equivalent to $B_{n+1} - B_{n+2}$, since the equivalence is set up by the partial isometry $V(A_n - A_{n+1})$, which can be seen just the same way as the equivalence $(A - A_1) \sim (B_1 - B_2)$ was shown to be set up by $V(A - A_1)$. Similarly, $B_n - B_{n+1}$ is equivalent to $A_{n+1} - A_{n+2}$, the equivalence given by $W(B_n - B_{n+1})$. So by the complete divisibility of $\sim$ we have

$$\sum_{n=0}^{\infty}(A_{2n} - A_{2n+1}) \sim \sum_{n=0}^{\infty}(B_{2n+1} - B_{2n+2}) \qquad (6.20)$$

$$\sum_{n=)}^{\infty}(A_{2n+1} - A_{2n+2}) \sim \sum_{n=0}^{\infty}(B_{2n} - B_{2n+1}) \qquad (6.21)$$

Since $A_\infty \sim B_\infty$ also holds, using the complete divisibility of $\sim$ again, we have

$$A = A_\infty + \sum_{n=0}^{\infty}(A_{2n} - A_{2n+1}) + \sum_{n=)}^{\infty}(A_{2n+1} - A_{2n+2})$$

$$\sim \sum_{n=0}^{\infty}(B_{2n+1} - B_{2n+2}) + \sum_{n=0}^{\infty}(B_{2n} - B_{2n+1})$$

$$= B$$

□

Proposition 6.12 (Comparison Theorem) *For any two $A, B \in \mathcal{P}(\mathcal{M})$ there is a projection $Z \in \mathcal{P}(\mathcal{M}) \cap \mathcal{P}(\mathcal{M})''$ such that $ZAZ \preceq ZBZ$ and $(I - Z)B(I - Z) \preceq (I - Z)A(I - Z)$.*

Proof: We prove the Comparison Theorem for factors only. What we must show is that if A, B are projections in a factor $\mathcal{M}$, then either $A \preceq B$ or $B \preceq A$. The proof makes use of the Zorn's lemma and the complete divisibility property of $\sim$. Let $\mathcal{F}$ be the family of sets $\{(A_i, B_i)\}$ of ordered pairs (A_i, B_i) such that A_i, B_i are equivalent subprojections of A and B respectively, and such that both $\{A_i\}$ and $\{B_i\}$ are orthogonal families. $\mathcal{F}$ is non-empty since it contains $\{(0,0)\}$. $\mathcal{F}$ can be partially ordered by the inclusion relation. The union of a totally ordered subset of $\mathcal{F}$ is an upper bound for the subset, and by Zorn's lemma $\mathcal{F}$ has a maximal element $\{(A_i, B_i)\}$ with respect to the inclusion relation. If $A - \sum_i A_i$ and $B - \sum_i B_i$ are non-zero, then they have equivalent non-zero subprojections A_0, B_0 by Proposition 6.9. Adjoining (A_0, B_0) to $\{(A_i, B_i)\}$ one obtains an element in $\mathcal{F}$ which is larger than $\{(A_i, B_i)\}$, and this contradicts the maximality of $\{(A_i, B_i)\}$. Thus either $A - \sum_i A_i$ or $B - \sum_i B_i$ is zero (possibly both).

By the complete divisibility $\sum_i A_i \sim \sum_i B_i$. Thus either A is equivalent to a subprojection of B, or B is equivalent to a subprojection of A. □

Proof: of Proposition 6.5 (Sketch). We first prove (iii) ⇒ (i): Let A be equivalent to the identity I i.e. $W^*W = I$ and $A = WW^*$ for some W. Then
$$T(WW^*) = T(W^*W) = I$$
hence
$$T(WW^* - I) = T(A - I) = 0$$
and property (v) of T implies $A = I$ hence $\mathcal{M}$ is finite.

(ii) ⇒ (i): The same as (iii) ⇒ (i).

The existence (and uniqueness) of both the center valued trace and of the family of traces is essentially a consequence of this lemma:

Lemma: If $\mathcal{M}$ is a finite von Neumann algebra and $\mathcal{Z}_\mathcal{M}$ is the center of $\mathcal{M}$, then a (not necessarily positive) linear functional ω defined on $\mathcal{M}$ is called *central functional* if
$$\omega(XY) = \omega(YX) \qquad \text{for all } X, Y \in \mathcal{M}$$

The statement of the Lemma is that any ultraweakly continuous ω linear functional defined on $\mathcal{Z}_\mathcal{M}$ can be extended uniquely to $\mathcal{M}$ to a central linear functional φ_ω, the extension φ_ω is bounded and ultraweakly continuous, positive if ω is positive and $\|\varphi\| = \|\omega\|$.

Proof of Lemma (Sketch)

The idea is to take any ultraweakly continuous extension of ω and consider its orbit under the shifts by unitary operators. The shifts act as a semigroup on the convex norm closure of the orbit, and the semigroup has a fixpoint, a state which is then unitary invariant, i.e. it is a trace. Let us see this in some more detail.

Let φ be an ultraweakly continuous extension of ω (this exists by the Hahn-Banach theorem). Let $U \in \mathcal{M}$ be unitary. Let us define the map $\mathcal{S}_U$ by:
$$(\mathcal{S}_U\varphi)(A) \equiv \varphi(U^*AU) \qquad (A \in \mathcal{M})$$
Furthermore, put
$$\mathcal{S}_\varphi \equiv \{\mathcal{S}_U\varphi \mid U \in \mathcal{M}\} \subseteq \mathcal{M}_*$$
and denote by $\mathcal{K}_\varphi$ the norm closure of the convex linear hull of $\mathcal{S}_\varphi$. Since $\mathcal{M} = \mathcal{M}_*{}^*$, the set $\mathcal{K}_\varphi$ is $\sigma(\mathcal{M}_*, \mathcal{M})$ closed (Mackey theorem). The key step – and this is where the finiteness of $\mathcal{M}$ is used – is that $\mathcal{K}_\varphi$ also is $\sigma(\mathcal{M}_*, \mathcal{M})$ compact. The compactness of $\mathcal{K}_\varphi$ implies that the semigroup $\mathcal{S}_U : \mathcal{K}_\varphi \to \mathcal{K}_\varphi$ ($U \in \mathcal{M}$ unitary) of affin maps has a fixpoint

(Ryll-Nardzewski fixpont theorem), i.e. there exists a fixpoint $\varphi_\omega \in \mathcal{S}_\varphi$, which means that $S_U \varphi_\omega = \varphi_\omega$ for all $U \in \mathcal{M}$ unitaries, implying that φ_ω is a central functional.

(i) $\Rightarrow$ (ii): Let ω be an ultraweakly continuous positive linear functional on $\mathcal{M}$. By the Lemma there exists the central positive functional $\tau_\omega \equiv \varphi_\omega$ (trace), which is finite. We omit the proof that the family $\{\tau_\omega \mid \omega \in \mathcal{M}_*^+\}$ is sufficient.

(i) $\Rightarrow$ (iii): We show first that a map $T: \mathcal{M} \to \mathcal{Z}_\mathcal{M}$ having the properties (i),(ii) and (iii) is unique if it exists. We must show that if T' is another map with the said properties, then $T'(X) - T(X) = 0$. To see this it suffices to show that $\omega(T'(X) - T(X)) = 0$ for every normal, linear functional ω on $\mathcal{Z}_\mathcal{M}$. But this is clear since both $\omega \circ T'$ and $\omega \circ T$ are extensions of ω (defined on $\mathcal{Z}_\mathcal{M}$) to a central functional, and by the Lemma the extension is unique, i.e. $\omega \circ T' = \omega \circ T$. We show now that there exists center valued trace. The map $f: (\mathcal{Z}_\mathcal{M})_* \to \mathcal{M}_*$ defined by $f(\omega) = \varphi_\omega$, which assigns the extension φ_ω to the normal functional ω is a linear isometry between $\mathcal{Z}_{\mathcal{M}*}$ and $\mathcal{M}_*$ by the Lemma. The dual f^* of f is, therefore a linear map from $\mathcal{M}_*^* = \mathcal{M}$ into $(\mathcal{Z}_{\mathcal{M}*})^* = \mathcal{Z}_\mathcal{M}$. Let $T \equiv f^*$, i.e. T is defined by $\omega(TX) = \varphi_\omega(X)$, ($\omega \in (\mathcal{Z}_\mathcal{M})_*$, $X \in \mathcal{M}$), where φ_ω is the extension of ω ensured by the Lemma. The properties (i),(ii),(iii) in the definition of T are thus immediate consequences of the fact that φ_ω is a central functional. To see $T(ZX) = ZT(X)$ where $X \in \mathcal{M}, Z \in \mathcal{Z}_\mathcal{M}$, it suffices to show that this last equality holds taking unitaries $U \in \mathcal{Z}_\mathcal{M}$ in place of Z. Let $T'(X) \equiv U^* T(UX)$. Then $T': \mathcal{M} \to \mathcal{Z}_\mathcal{M}$ is a map that satisfies the conditions (i)- (iii) in the definition of the center valued trace, hence $T' = T$, and so $U^* T(UX) = TX$, i.e. $T(UX) = UT(X)$. It remains to see that T has the property (v). Let $N \equiv \{X \in \mathcal{M} \mid T(X^*X) = 0\}$. Then the normality and the trace property of T implies that N is an ultraweakly closed two-sided ideal in $\mathcal{M}$. Every ultraweakly closed two-sided ideal in a von Neumann algebra $\mathcal{M}$ has the form $\mathcal{M}Z$, where $Z \in \mathcal{Z}_\mathcal{M} \cap N$ is a central projection. Thus $Z = ZT(I) = T(Z) = 0$ implying $Z = 0$, hence $N = \{0\}$. □

We next prove that the canonical center valued trace has the property (vi) of the dimension function.

Proposition 6.13 *Let A, B projections in a von Neumann algebra $\mathcal{M}$, and let T be the canonical center valued trace. Then the following are equivalent:*
(i) $A \preceq B$
(ii) $T(A) \leq T(B)$
In particular, it holds that

$$A \sim B \quad \text{if and only if} \quad T(A) = T(B) \tag{6.22}$$

Proof: Assume $A \preceq B$. Then $A = UU^*$ and $U^*U = B' \leq B$ with some partial isometry U, and by the trace property of T we have

$$T(A) = T(UU^*) = T(U^*U) = T(B')$$

Since T is positive we have

$$0 \leq T(B) - T(B') = T(B) - T(A)$$

i.e. $T(A) \leq T(B)$. If $A \not\preceq B$, then there exists a central projection Q by the Comparison Theorem such that $QB \prec QA$, and by the just shown (i) we have

$$T(QB) < T(QA)$$

which (in view of property (iv) of the center valued trace) implies

$$QT(B) < QT(A)$$

thus $T(A) \not\geq T(B)$. □

As it was seen the lattice of a type II_1 factor is modular. We mention finally that this remains true if the finite von Neumann algebra is not required to be factor:

Proposition 6.14 *If $\mathcal{M}$ is a finite von Neumann algebra, then $\mathcal{P}(\mathcal{M})$ is a modular lattice*

Proof: Let $A, B, C \in \mathcal{M}$ be projections in $\mathcal{M}$. One must show that the modularity equality (3.6) holds for A, B, C, i.e.

if $\quad A \leq B, \quad$ then $\quad A \vee (B \wedge C) = (A \vee C) \wedge B \quad (C \in \mathcal{P}(\mathcal{M}))$

Let $K \equiv A \vee (B \wedge C)$ and $H \equiv (A \vee C) \wedge B$. We show first by elementary reasoning that the following hold:

(i) $H \geq K$
(ii) $H \vee C = A \vee C = K \vee C$
(iii) $H \wedge C = C \wedge B = K \wedge C$

(i): Since $A \vee C \geq A$ and $A \vee C \geq C \geq C \wedge B$, it follows that

$$A \vee C \geq A \vee (C \wedge B)$$

furthermore $B \geq A$ and $B \geq C \wedge B$, thus (i) follows.
(ii): $(B \wedge (A \vee C)) \leq A \vee C$, hence

$$(B \wedge (A \vee C)) \vee C \leq (A \vee C) \vee C = A \vee C$$

however, one has $(A \vee C) \geq A$ and $B \geq A$, so $B \wedge (A \vee C) \geq A$, and it follows that
$$(B \wedge (A \vee C)) \vee C \geq A \vee C$$
Similarly: $A \vee (B \wedge C) \geq A$, thus
$$(A \vee (B \wedge C)) \vee C \geq A \vee C$$
Furthermore $A \leq A \vee C, (B \wedge C) \leq C$, and so
$$(A \vee (B \wedge C)) \leq (A \vee C) \vee C = A \vee C$$
(iii): $C \wedge B \leq B$ and $C \wedge B \leq C \leq A \vee C$, consequently
$$(C \wedge B) \leq (B \wedge (A \vee C)) \wedge C$$
Furthermore $B \geq B \wedge (A \vee C)$, and so
$$C \wedge B \geq (B \wedge (A \vee C)) \wedge C$$
One has $C \wedge B \leq A \vee (B \wedge C)$, hence
$$C \wedge B \leq (A \vee (B \wedge C)) \wedge C$$
Because of $B \geq B \wedge C$ and $B \geq A$ one has $B \geq A \vee (B \wedge C)$, thus
$$C \wedge B \geq (A \vee (B \wedge C)) \wedge C$$

H being finite, and (i) being the case, to see $H = K$ it is enough to show $H \sim K$. Using (ii)-(iii) and the paralellogramm rule (6.7) one can write
$$H - (C \wedge B) = H - (H \wedge C) \sim C \vee H - C =$$
$$= K \vee C - C \sim K - (K \wedge C) = K - (C \wedge B)$$
which implies $H \sim K$. □

6.4. Bibliographic notes

The theory of von Neumann algebras was established by J. von Neumann and Murray [104]. Originally, the von Neumann algebras were called "rings of operators", the name "von Neumann algebra" is due to Diximier. Historically, the study of Jordan algebras began with Jordan's 1933 papers [79], [80], and the first result on classification was obtained soon after by Jordan, von Neumann and Wigner in [81]. And it was von Neumann [170] who undertook an investigation of non-finite dimensional Jordan algebras; however, his work in this direction seems to be left unfinished by him. The dimension theory has remained unchanged since it was developed in [104]. The theory of operator algebras is described in many textbooks. The treatment in [158] and [144] is in the spirit of pure mathematics, the presentation in the two volume monograph [27] and [28] is motivated mainly by application of operator algebras in (quantum statistical) physics. The most comprehensive textbook on operator algebras is the recent multivolume book [82], [83]. For a brief review of von Neumann's legacy in the operator algebra theory see [111].

CHAPTER 7

The Birkhoff-von Neumann concept of quantum logic

The aim of this somewhat historical chapter is to describe the 1936 Birkhoff-von Neumann concept of quantum logic, which is markedly but subtly different from the standard view according to which quantum logic is an orthomodular (but not modular) lattice.

Quotations from the Birkhoff-von Neumann paper [21] presented in Section 7.1 show that for Birkhoff and von Neumann the modularity of quantum logic was an important property to require. Section 7.1 shows that Birkhoff – and especially von Neumann – insisted on modularity because von Neumann wanted quantum logic also to be a set of events on which a well-behaving non-commutative probability theory can be built, where "well-behaving probability theory" means a probability theory that allows an "a priori probability". Section 7.1.1 explains why for von Neumann the existence of an a priori probability was important, and Section 7.1.2 argues that von Neumann interpreted the trace on a type II_1 factor as the proper a priori probability on the proper quantum logic: the modular projection lattice of a type II_1 factor. The Birkhoff-von Neumann concept of quantum logic also makes it possible to interpret quantum probability in purely logical terms – this will be seen in Section 7.2.

7.1. Quantum logic as event structure of non-commutative probability

In their fundamental paper [21] Birkhoff and von Neumann postulated that the "quantum logic", i.e. the algebraic structure that should replace the logic (Boolean algebra) of a classical mechanical system

"*...has the same structure as an abstract projective geometry.* [their emphasis]" [21] ([178] p. 115) [1]

By an abstract projective geometry Birkhoff and von Neumann meant an orthocomplemented, *modular* lattice. As an example for such a lattice the

---

[1] When quoting from von Neumann's works that can be found in his Collected Works, we first refer to the original source (without giving page numbers) and this is followed (in parenthesis) by reference to the volume of von Neumann's Collected Works that contains the source and by the page number referring to the page in that volume on which the quoted text can be found.

projection lattice of a finite dimensional Hilbert space is cited by Birkhoff and von Neumann; they emphasize, however, that for quantum mechanics a "continuous-dimensional model" of projective geometry is preferred, and reference is made to von Neumann's work on continuous geometries (Footnote 33 in [21]). [2]

As remarked earlier, Birkhoff and von Neumann had been aware that the lattice of all projections on an infinite dimensional Hilbert space is not modular, thus for Birkhoff and von Neumann it was not the Hilbert lattice that represented the logic of a quantum system. What did Birkhoff and von Neumann then consider as the proper quantum logic and why did they insist on the modularity of the quantum logic?

The answer is, in short, that looking for a quantum logic Birkhoff and Neumann were in fact searching not only for a non-commutative (i.e. quantum) logic but also for a non-commutative generalization of classical probability theory. Now the elements in a Boolean algebra that represent the propositions associated with a classical mechanical system (in the sense described in Section 5.2) are considered in classical, commutative probability theory as *events*, and classical probability is viewed as a *normalized* measure μ on this Boolean algebra. The measure μ has the property

$$\mu(A) + \mu(B) = \mu(A \wedge B) + \mu(A \vee B) \qquad (7.1)$$

which is just one of the defining properties (property (ii)) of a dimension function on a lattice (see Definition 3.7). We have seen that the existence of such a *finite* dimension function implies the modularity of the lattice (Proposition 3.3). In conformity with the non-modularity of $\mathcal{P}(\mathcal{H})$ there is no normalized map on $\mathcal{P}(\mathcal{H})$ that would match the lattice operations in the required manner expressed by eq. (7.1). The absence of such a map was considered by von Neumann as a pathology of the usual Hilbert space quantum mechanics as a non-commutative probability theory. Birkhoff and von Neumann refer to this pathology in the form of calling the unique trace on a type II_1 factor as the "a priori thermodynamic weight" of states,[3] pointing out that the existence of such an a priori weight is related to the modularity of the lattice, i.e. there does not exist such an a priori weight on $\mathcal{B}(\mathcal{H})$ [21] ([178] p. 115). To understand the term "a priori thermodynamic weight of states" and its conceptual significance in connection with von Neumann's view of probability in quantum mechanics we must recall briefly

---

[2] This footnote also contains the somewhat surprising remark that these continuous geometries "... may be more suitable frame for quantum theory, than Hilbert space." ([178] p. 118)

[3] "...the presence of condition L5 [=modularity] is closely related to the existence of an 'a priori thermo-dynamic weight of states' " [21] ([178] p. 114-115)

von Neumann's concept of probability in quantum mechanics prior to his work on quantum logic and operator algebras.

7.1.1. DIGRESSION: VON NEUMANN'S CONCEPT OF PROBABILITY IN QUANTUM MECHANICS IN THE YEARS 1926-1932

Von Neumann's 1932 book [168], which summarizes the Hilbert space formalism of quantum mechanics is based on three papers von Neumann published in the year 1927 [165] [166] [167]. These three papers, all written while von Neumann was an assistant of Hilbert in Göttingen, were preceded by a joint work of D. Hilbert, L. Nordheim and J. von Neumann [68]. This latter paper was based on Hilbert's lectures in Göttingen on quantum mechanics, thus it is rather a document of Hilbert's views than that of von Neumann's ideas, but no attempt is made here to separate the three authors' views. In all these papers, as well as in the 1932 book, the notion of probability is of central importance, and in all these works difficulties in connection with probability in quantum mechanics arise. The difficulty in [68] is mainly of a technical nature, whereas in von Neumann's papers the mathematical part is impeccable, but conceptual problems appear instead.

The paper [68] attempts an axiomatic description of quantum mechanics and chooses as a primitive concept the *amplitude* $\phi(xy, F_1F_2)$ of the *density for relative probability*. This means that the quantity w defined by

$$\phi(xy, F_1F_2)\overline{\phi}(xy, F_1F_2) = w(xy, F_1, F_2) \tag{7.2}$$

is assumed to give the probability density for the probability that for a fixed value y of the quantity F_2 the value of the quantity F_1 lies between a and b, i.e. this probability is given by $\int_a^b w(xy, F_1, F_2) dx$.[4] In the Hilbert Nordheim von Neumann paper the amplitudes are identified with kernels of integral operators, and since the assumption is made that every operator is an integral operator, the Dirac function must be allowed as a kernel. In particular, the relative probability density for the probability that a quantity has the value x provided that the same quantity has the value y turns out to be given by $\delta(x - y)$. The interpretation given in the paper of this probability density is that "the probability relation between a quantity to itself distinguishes the value of the quantity infinitely sharply" [68] ([175] p. 119). Although this interpretation sounds very reasonable, the authors were fully aware that the Dirac function as they used it was not

---

[4]While $\int_a^b w(xy, F_1, F_2) dx$ thus gives the probabilities with respect to the condition that for a fixed value y of the quantity F_2, the value of the quantity F_1 lies between a and b, this is not the reason why w is called in [68] the *relative* probability density: 'relative' means here that w is not normalized, see [68], ([175] p. 107). We shall say more on von Neumann's distinction between relative and conditional probabilities below after equation (7.3).

mathematically legitimate (von Neumann never accepted the "improper" Dirac function, as it is well-known), and they clearly felt uneasy with every statement concerning any result involving operations with this illegitimate entity. This is very clear in the last few paragraphs of the paper, where the authors support their decision not to elaborate further on some natural questions by pointing out that it is unclear whether/when the formal mathematical operations are meaningful; furthermore, explicit reference is made in this part of the paper to von Neumann's to-be-published work on the mathematical foundations of quantum mechanics, where a mathematically unobjectionable formalism is said to be worked out.

It turned out, however, that a new, conceptual difficulty arises even if the problematic Dirac function is eliminated from the formalism: infinite probabilities appear in the theory.

In the paper on the mathematical foundation of quantum mechanics [165], in which quantum mechanics (and the eigenvalue problem for possibly unbounded selfadjoint operators in particular) is worked out without making use of the Dirac function, the heart of the whole theory, the "statistical Ansatz", is formulated in Section XIII ([165] ([175] p. 195). This all-important Ansatz states that the relative probability that the values of the pairwise commuting quantities S_i lie in the intervals I_i if the values of the pairwise commuting quantities R_j lie in the intervals J_j are given by

$$Tr\Big(E_1(I_1)E_2(I_2)\ldots E_n(I_n)F_1(J_1)F_2(J_2)\ldots F_m(J_m)\Big) \qquad (7.3)$$

$E_i(I_i), F_j(J_j)$ being the spectral projections (belonging to the respective intervals) of the corresponding operators S_i and R_j.[5] The reason why von Neumann calls the probability in (7.3) *relative* is not that (7.3) gives the probability of events $E_i(I_i)$ relative to the condition that the values of R_j lie in the intervals J_j, i.e. he does not interpret 'relative' as 'conditional'. As we shall see below, apart from the so-called 'a priori' probability, for him every probability is 'conditional' in the sense that it is to be computed as a relative frequency in an ensemble selected from the a priori ensemble by some well-defined condition (such as the condition that the values of R_j lie in the intervals J_j for instance). 'Relative' means for von Neumann that the probability in (7.3) is not normalized in general (von Neumann calls the correctly normalized probabilities 'absolute'), see ([168] p. 310-311).

As von Neumann stresses, the distinction between relative and absolute probability in this sense is not significant as long as the probability in (7.3)

---

[5]Note that von Neumann did not use the notation in (7.3): using the trace Tr is a later development. In von Neumann's notation $Tr(E_iF_j)$ reads $[E_i, F_j]$. Von Neumann's notation is somewhat unfortunate from today's point of view, since today one would read $[E_i, F_j]$) as the commutator of the operators involved.

is finite, for one can then normalize it. However, von Neumann realizes that the relative probability (7.3) can be infinite. This happens if any of I_i or of J_j contains parts of the continuous spectrum of R_i or S_j. And the a priori probabilities can also be infinite, that is the probabilities obtained from (7.3) by substituting $F_1(J_1)F_2(J_2)\ldots F_m(J_m) = I$ into (7.3) [165], ([175] p. 194,197).

Von Neumann tries to justify the usage of infinite probability in [166] and in [168] (p. 165) by referring to the following example: If a quantity represented by a real valued function f can take its value anywhere in the real line with equal probability, then – if the probability were normalized – the probability that f takes its value in any interval d would be equal to zero, "... but the equal probability of all places is not expressed in this way, but rather by the fact that two finite intervals have as their probability ratio the quotient of their length. Since 0/0 has no meaning, this can be expressed only if we introduce their lengths as relative probabilities – the relative total probability will then of course be ∞." ([168] p. 310)

This example only explains why one should allow infinite *total* (relative) probabilities, but it does not make clear what sense to make out of relative probabilities that *themselves*, not only their sum, are infinite. This problem remains, and it becomes even more serious if considered in connection with von Neumann's attempt in [166] to "work out inductively" [166] ([175] p. 209), which meant for von Neumann to *derive* the statistical formula (7.3) from the basic principles of probability theory.

The starting point of the derivation in [166] is the assumption of an *elementary unordered ensemble* ("elementar ungeordnete Gesamtheit"). Von Neumann also calls this ensemble "fundamentale Gesamtheit" in [167] ([175] p. 232), and the terminology "ensemble corresponding to 'infinite temperature'" and "absoluter Gleichgewichtszustand" also appear in [167] ([175] p. 238). This is the *a priori* ensemble, of which one does not have any specific knowledge. Every system of which one knows more is obtained from this ensemble by selection: one checks the presence of a certain property P (e.g. that quantity Q has its value in the set I) on every element of the a priori ensemble and collects into a new ensemble those elements which have the property P. This new ensemble E is the one on which one has to compute the relative probabilities (relative to the condition P) by checking again the presence or absence of a certain property, collecting those elements that have the property and computing the frequency of these in E. Identifying ensembles with expectation-value-assignments and assuming the formalism of quantum mechanics (but not (7.3)), von Neumann shows that each ensemble can be described by a "statistical operator", a positive, non-zero operator U, where the description is meant in the sense that the expectation values determined by the ensemble corresponding to U are given by $Tr(UQ)$. The statistical operator of the a priori ensemble is the unit operator I.

Note that the formula $Tr(UQ)$ is not yet what von Neumann wants to have: the aim is to derive the statistical Ansatz of quantum mechanics (i.e. (7.3)) that yields the relative probabilities. To arrive at that formula von Neumann carries out what in today's terminology is called "statistical inference": Suppose that we only know of a system S that the values of the pairwise commuting quantities R_j lie in the intervals J_j. What statistical operator for the ensemble should we infer from this knowledge? – asks von Neumann. *Assuming* that it was the a priori ensemble on which we checked that the quantities R_j lie in the intervals J_j, and have collected those members of *this ensemble* on which this property was found present into a new ensemble E', von Neumann shows that the statistical operator is $F_1(J_1)F_2(J_2)\ldots F_m(J_m)$.

Von Neumann emphasizes that his derivation of the wanted statistical operator $F_1(J_1)F_2(J_2)\ldots F_m(J_m)$ crucially depends on how the ensemble E' was produced, [166] ([175] p. 228): since E' was assumed to be obtained from the a priori ensemble E, the whole derivation and the result depend on the a priori ensemble. The significance of this point is made particularly clear in [168], where the statistical inference of [166] is reproduced in Chapter IV: That we have the outcomes of the measurements of R_j in the ensemble E' to lie in J_j

> '... can be attributed ... to the fact that originally a large ensemble E was given in which the measurements were carried out, and then those elements for which the desired results occurred were collected into a new ensemble. This is then E'. Of course everything depends on how E is chosen. This initial ensemble gives, so to speak, the a priori probabilities of the individual states of the system S. The whole state of affairs is well-known from the general theory of probability: to be able to conclude from the results of the measurements to the states, i.e. from effect to cause, i.e. to be able to calculate a posteriori probabilities, we must know the a priori probabilities.' ([168] p. 338)

On page 183 in [168] von Neumann once again stresses the significance of the a priori ensemble E in the derivation of (7.3) and adds

> 'In $E = I$ therefore, all possible states are in the highest possible degree of equilibrium, and no measuring action can alter this. For each orthonormal set $\phi_1, \phi_2 \ldots$,

$$E = I = \sum_{n=1}^{\infty} P_{[\phi_n]}$$

i.e. the 1 : 1 : ... mixture[6] of all states $\phi_1, \phi_2 \ldots$. From this we learn that $E = I$ corresponds to the assumption in the older quantum theory of the "a priori equal probability of all simple quantum orbits". It will also play an important role in our thermodynamical considerations, to which the next sections are devoted.' ([168] p. 346)

Characteristically, those paragraphs, as well as von Neumann paper [167], which those paragraphs are based on, have to make the *extra* assumption from beginning through the end that the statistical operator is *normalized*, i.e. that there are no infinite probabilities.[7] It is in this paper [167] that positive, normalized, trace class operators ("density matrices (operators)" as they have become to be called) are consequently considered as the mathematical objects yielding the expectation values of quantum observables.

Summing up: Von Neumann regarded the density matrices not as quantum states but as "statistical operators" describing *ensembles* of systems prepared in different quantum states. More specifically, a density matrix $\rho = \sum_i \lambda_i P_i$ is viewed in [168] and especially in [166] and in [167] as a statistical operator describing an ensemble of systems whose members are prepared in the quantum states P_i (P_i being the one dimensional projection determined by a single state vector η_i) and the fraction of the ensemble composed of systems prepared in state P_i is λ_i. The expectation value of an observable Q in the statistical ensemble described by ρ is given by

$$\langle Q \rangle_\rho = \text{Tr}(\rho Q) = \sum_j \langle \eta_j, \rho Q \eta_j \rangle \tag{7.4}$$

In von Neumann's interpretation of quantum probability, a distinguished role is played by the special *a priori ensemble*. This is the completely unselected ensemble from which all other, selected ensembles are obtained, and von Neumann considered assuming the existence of this ensemble indispensable to derive the formula yielding the quantum mechanical (relative) probabilities. In this completely non-selected ensemble, systems prepared in all conceivable states are supposed to be present with "equal relative frequency". The statistical operator of this fundamental ensemble is given by the identity operator I. The a priori weight of the quantum states is just 1 and this weight is equal to the trace (=dimension) of the one dimensional projection determined by the quantum state. Thus, given any

---

[6] The English translation uses 'superposition'. The original text says 'Gemisch', and it is clear that von Neumann means what one today calls 'mixture'.

[7] See p. 192, 208 in [168]. Discussing the entropy formula $S = -NkSpur(U\ln U)$ von Neumann writes: 'Here we have to assume $Spur(U) = 1$, i.e. that the expectation values of **S** [=the system] are absolute expectation values; contrary to [165], this assumption cannot be avoided here. ' [167], ([175] p. 240), my translation.

projection P, its trace (= number of pairwise orthogonal one dimensional projections P majorizes) is just its a priori weight (probability) in the statistical operator I of the elementary unordered, unselected ensemble. Viewed from this perspective, the formula (7.4) can be read in this way: Based on some preparation procedure, we know that a sub-ensemble has been selected from the completely unselected one, a sub-ensemble which consists of systems prepared in states P_i whose a priori weight (=1) is replaced by the "a posteriori weight" λ_i (and the "a posteriori weight" of all other states is just zero).

Von Neumann's interpretation and derivation has serious conceptual difficulties: The main problem is that the statistical operator I is not normalized, i.e. its trace is infinite, as is the trace of any infinite dimensional projection; in other words, there exists no non-zero, additive, finite probability on the set $\mathcal{P}(\mathcal{H})$ of all projections which would give the a priori, uniform probability on the set of one dimensional projections. A consequence of the a priori probabilities ("weights") not being finite is that the "a priori weights" *cannot* in fact be considered as relative frequencies (which are always ≤ 1) in any ensemble. A further difficulty is that the a priori, "elementar ungeordnet" ensemble is not an ensemble in which relative frequencies can be computed at all, since this ensemble contains, by definition, an uncountable number of elements (since the number of different projections on $\mathcal{H}$ is not countable); thus, strictly speaking, the notion of relative frequency does not make sense in such "ensembles".

It will be argued below in detail that von Neumann was bothered by the infinity of the trace, but there is also circumstantial evidence that von Neumann was aware of the internal inconsistency of his notion of a priori ensemble in quantum mechanics: when defining this ensemble in [166] von Neumann writes:

> 'The basis of a statistical investigation is always that one has an "elementary unordered ensemble" $\{S_1, S_2, \ldots\}$, in which "all conceivable states of the system S occur with equal relative frequency"; one must associate the distribution of values on this ensemble with those systems S, states of which one does not have any specific knowledge.' [166], ([175] p. 210), my translation.

Notice the quotation marks in the above formulation of von Neumann. Since he is not actually quoting from any source, it is reasonable to interpret the quotation marks as conveying the message that what they enclose is not to be taken literrally. This interpretation is further supported by the fact that in Chapter IV in [168], in which the paper [166] is reproduced, one nowhere finds the above cited specification of the a priori ensemble: ensembles in [168] are always explicitly assumed to have an arbitrary large, *finite* number of elements. While relative frequencies certainly exist in such ensembles, it is unclear how they can serve to determine the a priori probabilities of *all conceivable* states.

One reaction to the conceptual difficulty related to the existence of infinite probabilities can be to take the position that the a priori weights are not probabilities in the sense of relative frequency. Taking this position means in effect that one abandons the relative frequency interpretation of probability in favor of an interpretation which could in principle be compatible with infinite probabilities. For von Neumann this option was probably out of the question: in his paper [166] von Neumann speaks of a frequency interpreted probability as *the* (i.e. unique) theory of probability. His view of probability was clearly shaped under the influence of von Mises' relative frequency interpretation. Although there is no mention of von Mises' name in the three fundamental papers [165], [166], [167], von Neumann probably had known von Mises' frequency interpretation already in 1926-1927, and he refers explicitly to von Mises' 1928 book on probability [100] in the footnote 156 in [168]. In this footnote von Neumann identifies what he takes as the quantum mechanical ensembles with what von Mises calls "Kollektive". However, in von Mises' theory a Kollektive is always countable; furthermore, von Mises explicitly says that "The purpose of the calculus of probability, strictly speaking, consists exclusively in the calculation of probability distributions in new collectives derived from given distributions in certain initial collectives." ([100] p. 221) Accordingly, in von Mises' view there is no a priori knowledge of probabilities.

Another option is to abandon the idea that the statistical operator I gives the a priori uniform probability. As we have seen, von Neumann himself made clear how indispensable the assumption of the a priori probabilities/weights are if one wants to be able to derive the statistical formula (7.3) via statistical inference. So von Neumann could not do without the a priori probabilities as long as he wanted to found the statistical Ansatz. Facing the clash between the necessary but infinite a priori probability and the frequency interpretation of probability, and not wanting or being able to abandon either the a priori probability or the frequency interpretation, von Neumann was left with one option only, which is a radical one: To consider the appearance of infinite, not normalizable "a priori probabilities" as a pathology (from the point of view of probability theory) of the Hilbert space quantum mechanics, and to try to work out a well-behaving non-commutative probability theory, one in which there exists (normalized) a priori probability, an "a priori thermodynamic weight of states".

7.1.2. BACK TO THE TYPE II$_1$ FACTOR

This is a program which goes beyond the Hilbert space quantum mechanics, and which von Neumann succesfuly completed while working on quantum logic in 1936 by classifying the factors and by discovering the type II$_1$ factor: there exists such an a priori probability on the lattice of a type II$_1$ factor: it is given (uniquely) by the trace.

That von Neumann did indeed become unfaithful to the Hilbert space formalism is especially clear from his famous letter to Birkhoff, where he writes:

> "I would like to make a confession which may seem immoral: I do not believe absolutely in Hilbert space any more. After all Hilbert space (as far as quantum mechanical things are concerned) was obtained by generalizing Euclidean space, footing on the principle of 'conserving the validity of all formal rules'.... Now we begin to believe that it is not the *vectors* which matter, but the lattice of all linear (closed) subspaces. Because: 1) The vectors ought to represent the physical *states*, but they do it redundantly, up to a complex factor only, 2) and besides, the states are merely a derived notion, the primitive (phenomenologically given) notion being the qualities which correspond to the *linear closed subspaces*. But if we wish to generalize the lattice of all linear closed subspaces from a Euclidean space to infinitely many dimensions, then one does not obtain Hilbert space, but that configuration which Murray and I called 'case II$_1$'. (The lattice of all linear closed subspaces of Hilbert space is our 'case I$_\infty$'." [169]

It is the uniquely determined trace τ – or equivalently the dimension function d on a type II$_1$ factor – which von Neumann interpreted as the proper a priori probability in quantum mechanics. In the Introduction to the first paper on the rings of operators, where Murray and von Neumann give the quantum mechanical motivation of the whole project of the investigation of the different types of von Neumann algebras, the authors discuss the significance of the type II$_1$ case and write

> "Considering the immediate applicability of $T(A)$ [$= \tau(A)$] to quantum mechanics it is the 'a priori' expectation values of the observable A, which is correctly normalized here, but cannot be in I$_\infty$..." [104] ([177] p. 11)

Note that the a priori probability on a type II$_1$ factor is not a uniform probability measure on the set of events (i.e. on the lattice): If $\{A_i\}$ is an infinite set of disjoint events, then obviously there cannot exist a non-zero, additive, normalized, uniform measure p on $\{A_i\}$ (i.e. a p such that $p(A_i) = c > 0$ for all A_i), no matter whether the events form a distributive lattice or not. "A priori", therefore, acquires a new meaning: it reflects the symmetry of the system: The dimension function (=probability) on the lattice of a type II_1 factor comes from a trace. Now, as we have seen, a trace is just the *unique* (positive, linear, normalized) functional that is invariant with respect to all unitary transformations. Since the physical symmetries of the system are generally expressed as representations of the symmetry group on the algebra of observables by unitaries, the existence of a unique trace means physically that the probability is determined uniquely as the only (positive, linear) assignment of values in $[0, 1]$ to the events that is invariant with respect to any conceivable symmetry.

The idea that symmetry and a priori probability are closely related is not new: in the naive, common sense reasoning concerning the chances of throwing a certain number with an ordinary *unloaded* die one in fact identifies the unloadedness of the die with its perfect symmetry, and infers form this the equality of the probabilites of the events. This inference is made without making any actual throw, so the uniform probabilities appear as the a priori probabilities of the events. In the presence of an infinite number of events the uniformity of the a priori probabilities cannot be preserved. This is a well-known problem already in classical probability theory, where, too, one tries to infer the a priori probabilities from symmetry considerations. The difficulty is that a classical event structure does not by itself determine a natural symmetry group. In contrast to this, a quantum event structure does, which shows a greater coherence of quantum logic.

The Birkhoff-von Neumann remark on the existence of an "a priori thermodynamic weight of states",[8] though quite obscure and hardly understandable in its original context, should now make perfect sense: for von Neumann the usual trace Tr on an infinite dimensional Hilbert space gives the a priori thermodynamic weight of states in the sense discussed in Section 7.1, but this trace does not exist as a finite quantity. If we want to have a finite "a priori thermodynamical weight", i.e. a finite a priori probability, then the lattice must be modular (Proposition 3.3), and one can indeed have a non-commutative algebra of quantum observables whose lattice is of this type: the typical example is the II_1 von Neumann factor algebra, together with its unique trace.

7.2. Probability is logical

Furthermore, the concept of quantum logic as the projection lattice of a type II_1 factor also made it possible to give a purely logical interpretation to the associated non-commutative probability, and this is another reason

---

[8] See the footnote 3.

why Birkhoff – and especially von Neumann – found the lattices in question significant: The unitaries, with respect to which the trace (probability) is invariant, define isomorphisms of the lattice i.e. of the logic of the system, and so the probability appears as being fixed once the logic of the system is fixed. In von Neumann's words:

> "Essentially if a state of a system is given by one vector, the transition probability in another state is the inner product of the two which is the square of the cosine of the angle between them. In other words, probability corresponds precisely to introducing the angles geometrically. Furthermore, there is only one way to introduce it. The more so because in the quantum mechanical machinery the negation of a statement, so the negation of a statement which is represented by a linear set of vectors, corresponds to the orthogonal complement of this linear space. And therefore, as soon as you have introduced into the projective geometry the ordinary machinery of logics, you must have introduced the concept of orthogonality. ... In order to have probability all you need is a concept of all angles, I mean angles other than 90°. Now it is perfectly quite true that in geometry, as soon as you can define the right angle, you can define all angles. Another way to put it is that if you take the case of an orthogonal space, those mappings of this space on itself, which leave orthogonality intact, leave all angles intact, in other words, in those systems which can be used as models of the logical background for quantum theory, it is true that as soon as all the ordinary concepts of logic are fixed under some isomorphic transformation, all of probability theory is already fixed."

(Unpublished, quoted in [31].)

Von Neumann also considered the theory of von Neumann lattices as a kind of "quantum set theory". The analogy of the equivalence $\sim$ between projections with the Cantorian notion of cardinality is striking, and von Neumann explicitly says

> "... the whole algorithm of Cantor theory is such that most of it goes over in this case. One can prove various theorems on the additivity of equivalence and the transitivity of equivalence, which one would normally expect, so that one can introduce a theory of alephs here, just as in set theory. ... I may call this dimension since for all matrices of the ordinary space, it is nothing else but dimension."

(Unpublished, quoted in [31].)

> "One can prove most of the Cantoreal properties of finite and infinite, and, finally, one can prove that given a Hilbert space and a ring in it, a simple ring in it [i.e. a factor], either all linear sets except the null set are infinite (in which case this concept of alephs gives you nothing

new), or else the dimensions, the equivalence classes, behave exactly like numbers and there are two qualitatively different cases. The dimensions either behave like integers, or else they behave like all real numbers. There are two subcases, namely there is either a finite top or there is not."

(Unpublished, quoted in [31])

Von Neumann calls a complete lattice $\mathcal{L}$ "continuous geometry" if the following hold

(i) $\mathcal{L}$ is complemented, i.e. for every $A \in \mathcal{L}$ there exists an A' such that $A \vee A' = I$ and $A \wedge A' = 0$
(ii) $\mathcal{L}$ is modular
(iii) $\mathcal{L}$ is "continuous", i.e. if $\mathcal{S} \subseteq \mathcal{L}$ is a partially ordered set in $\mathcal{L}$ then for any $A \in \mathcal{L}$ it holds that

$$A \wedge (\vee \{S \quad S \in \mathcal{S}\} = \vee \{A \wedge S \quad S \in \mathcal{S}\}$$
$$A \vee (\wedge \{S \quad S \in \mathcal{S}\} = \wedge \{A \vee S \quad S \in \mathcal{S}\}$$

A key concept in the analysis of continuos geometries is the notion of "perspectivity": the elements A and B in $\mathcal{L}$ are called perspective if they have a common complement, i.e. there exists an $S \in \mathcal{L}$ such that $A \vee S = B \vee S = I$ and $A \wedge S = B \wedge S = 0$. The perspectivity relation is the lattice theoretic counterpart of the $\sim$ equivalence relation between projections. The lattice theoretic counterpart of the center of a von Neumann algebra is the center of the lattice: it is the set of elements that have one complement only, and the lattice is called *irreducible* if its center contains only the elements $0, I$.

Von Neumann could prove that on an irreducible continuous geometry there exists a d dimension function taking on each value in the interval $[0, 1]$, i.e. a function $d: \mathcal{L} \to [0, 1]$ having the properties
(i) $d(I) = 1$, $d(A) = 0$ if and only if $A = 0$;
(ii) if $A \wedge B = 0$ then $d(A \vee B) = d(A) + d(B)$;
(iii) if A, B are perspective, then and only then $d(A) = d(B)$.

The difficult part of the proof of properties of d is to show that the perspectivity relation is transitive. Kaplansky proved in 1955 that if a lattice is not only complemented but *ortho*complemented, then the modularity of the lattice implies that the continuity property (7.5) holds, i.e. the lattice is a continuous geometry. Thus a continuous geometry is just an orthocomplemented modular lattice, and the projection lattice of a type $\mathbf{II}_1$ factor is a continuos geometry. The continuous geometries and their dimension theory can be considered as the lattice theoretic generalization of the lattices of type $\mathbf{II}_1$ factors.

Von Neumann could also prove, however, that in fact the von Neumann lattices form a "very large class" of continuous geometries: every continous geometry on which a suitably defined transition probability function exist is isomorphic to a von Neumann lattice.

We conclude this Chapter with three remarks. The first is that the existence of the trace (a priori probability) was not the only reason why von Neumann preferred the type II_1 structure to the I_∞ case (i.e. to the Hilbert space formalism). For instance the nice behavior of the unbounded operators affiliated with a II_1 von Neumann algebra – as opposed to the very pathological properties of the set of all unbounded operators on a Hilbert space – was another reason (see [104]).

The second remark is that the development of post-Hilbert-space quantum theory did not completely confirm von Neumann's expectation regarding the privileged role in quantum theory of the type II_1 structure. It turned out that even the most "pathological", type III von Neumann algebras occur naturally in quantum mechanical applications, for instance in relativistic quantum field theory [56] (see the Chapter 10).

The third remark is of general character and concerns the physics community's perception and evaluation of von Neumann's achivements in quantum mechanics. The attitude of the physics community towards von Neumann's work has remained ambivalent: while von Neumann's work is appreciated as an outstanding intellectual achievement, in fact his work is looked upon by physicists as a luxury, as an example of striving for mathematical exactness for exactness' own sake. On the other hand, the precise content of (and physical motovation for) von Neumann's admittedly rather mathematical results seems to be not widely known. Hopefully the present chapter also shows that the common perception of von Neumann as a physicist is onesided: the moral of the presented story of von Neumann's intellectual move from the Hilbert space formalism towards the type II_1 (and even more general) algebras is that what drove him was not the desire to have a mathematically unobjectionable theory – there was nothing wrong with Hilbert space formalism as a mathematical theory. What von Neumann wanted was conceptual understanding. He was ready to leave behind any mathematical theory – however beautiful in itself – to achieve that.

7.3. Bibliographic notes

While it is well-known that Birkhoff and von Neumann postulated the modularity of quantum logic despite the fact that they knew that the quantum logic one can extract from the Hilbert space formalism is non-modular, the question why they did so, seems to have been completely neglected both by quantum logicians and historians of quantum mechanics. The only two papers known to the present author that discuss this issue are Bub's works [31], [32]. The non-atomic character of the lattice of a type II_1 factor is analyzed in [31] also from the point of view of the measurement problem in quantum mechanics. The historical reconstructions in Sections 7.1.1 and 7.1.2 are based on [133]. The problem of a priori probabilities in classical probability theory and their relation to symmetry is analyzed in Section 3 of Chapter 3 in Van Fraassen book [163]. A particularly outspoken defender of the view that a priori probabilities should be determined in classical probability theory by symmetry considerations is E.T. Jaynes, see [77]. The theory of abstract orthomodular lattices originates in the theory of operator algebras. The line of development *von Neumann algebras* → *continuous geometries* → *orthomoduar lattices* and the connections between the theory of von Neumann algebras and lattice theory is described in the non-technical review [69].

CHAPTER 8

Quantum conditional and quantum conditional probability

It is well known that in the classical, elementary propositional logic one can define the classical conditional connective $A \to B = A^\perp \vee B$. This conditional, which is also known as "material implication", or "horseshoe", describes certain features of the inference "if A...then B". In this chapter the basic properties of the so-called "quantum conditional" are described. The quantum conditional is the analogue of the classical conditional connective, it is defined in the Hilbert lattice $\mathcal{P}(\mathcal{H})$ (or more generally in any orthomodular lattice) in terms of the lattice operations $\wedge, \vee$ and $\perp$ in $\mathcal{P}(\mathcal{H})$, and it reflects certain features of "if...then" inference. The semantic content of the quantum conditional is given by the "minimal implicative criteria", which the quantum conditional is required to satisfy. Section 8.1 describes these minimal implicative criteria. It turns out that there exist three conditionals satisfying the minimal implicative criteria. Each of these conditionals is a natural non-commutative generalization of the classical horseshoe, and each of them violates some conditions that the classical conditional satisfies (Propositions 8.2 and 8.4). Proposition 8.3 is the main result in Section 8.1, it tells us that the orthomodularity of $\mathcal{P}(\mathcal{H})$ is equivalent to the existence of a unique quantum conditional that satisfies the natural weakening of the classical implication condition. A further characteristic property of the quantum conditional is that it is a counterfactual conditional (Proposition 8.5). The main idea of the possible world semantics of counterfactuals is also recalled in Section 8.1.

Having recalled the notion of conditional probability (and the theory of conditionalization in general) together with the problem of statistical inference in Section 8.2, it is shown in Section 8.3 that Stalnaker's Thesis ("probability of conditional=conditional probability") is violated in quantum logic if the quantum conditional probabilities are given by a conditional expectation and the conditional is a quantum conditional satisfying the minimal implicative criteria.

8.1. Minimal implicative criteria and quantum conditional

It is important to keep in mind the distinction between an "if – then" implication *relation* and an "if – then" implication *connective*. The partial ordering relation $\leq$ defined on $\mathcal{P}(\mathcal{H})$ is a semantic relation: as we have seen in Chapter 5, $\leq$ expresses semantic entailment: the meaning of $A \leq B$ was that "if A is true then B is true"; the relation $\leq$ *mentions* A and B.

In contrast, a conditional operation $\Rightarrow$ is defined (in terms of the other logical operations of $\wedge$, $\vee$ etc.) in the language itself, and *using* A and B, the conditional $\Rightarrow$ forms a third element $A \Rightarrow B$ in the language. The element $A \Rightarrow B$ is not arbitrarily formed from A, B, of course, but in such a way that $\Rightarrow$ satisfy certain criteria with respect to the semantic entailment relation $\leq$.

It is a minimal requirement for instance that if it is the case that "if A is true then B is also true", then $A \Rightarrow B$ be valid, i.e. one requires the "law of entailment" to hold:

(E)   if $A \leq B$, then $(A \Rightarrow B) = I$

One expects of a conditional furthermore to satisfy the modus ponens: If A is true and A implies B, then B also is true. Formally:

(MP) $A \wedge (A \Rightarrow B) \leq B$

In an orthocomplemented lattice it is natural furthermore to require that the condtional be related to the "not" operation, i.e. one demands that if "not-A" is true and B implies A, then "not-B" is true. Formally:

(MT) $B^\perp \wedge (A \Rightarrow B) \leq A^\perp$

Calling E, MP and MT "minimal implicative criteria" (MIC), there exist three conditional operations which can be written as lattice polinoms and which satisfy MIC. These three conditionals are the following:

$$\begin{aligned} A \Rightarrow_Q B &= A^\perp \vee (A \wedge B) \\ A \Rightarrow_2 B &= (A^\perp \wedge B^\perp) \vee B \\ A \Rightarrow_3 B &= (A \wedge B) \vee (A^\perp \wedge B) \vee (A^\perp \wedge B^\perp) \end{aligned}$$

Each of these conditionals is a generalization of the classical material implication

$$(A \Rightarrow B) = A^\perp \vee B$$

in the sense that each of the three reduces to the classical conditional if restricted to a Boolean sublogic.

The non-classical feature of $\Rightarrow_Q, \Rightarrow_i$ (i=2,3) manifests in the fact that theese conditionals violate certain implicative criteria that are satisfied by the classical conditional.

Such a classical implicative criteria is the "law of exportation":

(EXP)   if $(A \wedge C) \leq B$ then $C \leq (A \Rightarrow B)$

(If the truth of A and C imply the truth of B, then the truth of C implies that "A implies B" is true.)

(EXP) and (MIC) together are equivalent to the "classical implicative criteria":

(CIC) $\quad\quad\quad A \wedge C \leq B \quad \text{if and only if} \quad C \leq (A \Rightarrow B)$

Proposition 8.1 *Let $\mathcal{L}$ be an orthocomplemented lattice. If there exists a conditional $\Rightarrow$ in $\mathcal{L}$ that satisfies the classical implicative criteria (CIC), then $\mathcal{L}$ is distributive and the conditional is identical to the classical material implication.*

Proof: Let $A, B, C \in \mathcal{L}$ be three elements in the orthocomplemented lattice $\mathcal{L}$. Then

$$\begin{aligned} (A \wedge C) &\leq (A \wedge C) \vee (B \wedge C) \\ (B \wedge C) &\leq (A \wedge C) \vee (B \wedge C) \end{aligned}$$

thus if $\Rightarrow$ is a conditional satisfying (CIC), then

$$\begin{aligned} A &\leq (C \Rightarrow ((A \wedge C) \vee (B \wedge C))) \\ B &\leq (C \Rightarrow ((A \wedge C) \vee (B \wedge C))) \end{aligned}$$

which implies that

$$(A \vee B) \leq (C \Rightarrow ((A \wedge C) \vee (B \wedge C)))$$

Applying (CIC) again we obtain

$$((A \vee B) \wedge C) \leq (A \wedge C) \vee (B \wedge C)$$

thus the lattice is distributive. In a distributive lattice the material implication is the only conditional satisfying the minimal implicative criteria. □

Since $\mathcal{P}(\mathcal{H})$ is not distributive it follows that

Proposition 8.2 *None of the three conditionals $\Rightarrow_Q, \Rightarrow_2, \Rightarrow_3$ satisfies the classical implicative criteria (CIC); hence each of them violates the law of exportation.*

There exists a weakening of (CIC) that can be satisfied by a quantum conditional: the so-called "quasi implicative condition":

(QIC) $\quad\quad \text{if } (A \wedge C) \leq B \text{ then } A^\perp \vee (A \wedge C) \leq (A \Rightarrow B)$

Proposition 8.3 (Mittelstaedt Theorem) $\Rightarrow_Q$ *satisfies the condition* (QIF); *furthermore, if in an orthocomplemented lattice* $\mathcal{L}$ *there exits a conditional* $\Rightarrow$ *which satisfies the minimal implicative criteria and the condition* (QIC), *then* $\mathcal{L}$ *is an orthomodular lattice and* $\Rightarrow$ *is unique, and coincides with* $\Rightarrow_Q$.

(The conditional $\Rightarrow_Q$ is often called Mittelstaedt conditional, and it can be found in the literature also under the name "Sasaki hook". Mittelstaedt himself refers to it as "material quasi-implication".)

Proof: If $A \wedge C \leq B$ then

$$A \wedge C \leq A \wedge B$$

thus

$$A^\perp \vee (A \wedge C) \leq A^\perp \vee (A \wedge B) = (A \Rightarrow_Q B)$$

and so $\Rightarrow_Q$ satisfies (QIC). We show that $\Rightarrow_Q$ is unique. Assume that $\hookrightarrow_Q$ also satisfies the minimal implicative criteria and (QIC). Putting $C = B$ into (QIC) we obtain:

$$A^\perp \vee (A \wedge B) \leq (A \hookrightarrow_Q B)$$

and so

$$(A \Rightarrow_Q B) \leq (A \hookrightarrow_Q B) \tag{8.1}$$

Since $\hookrightarrow_Q$ satisfies modus tollens

$$A \wedge (A \hookrightarrow_Q B) \leq B$$

it follows that

$$A \wedge (A \hookrightarrow_Q B) \leq A \wedge B \leq A^\perp \vee (A \wedge B) = (A \Rightarrow_Q B)$$

therefore

$$A^\perp \vee (A \wedge (A \hookrightarrow_Q B)) \leq (A \Rightarrow_Q B) \tag{8.2}$$

Becuse of (8.1) it holds that

$$A^\perp \leq (A \hookrightarrow_Q B)$$

so we can apply the orthomodularity relationon to the left hand side of (8.2) and we obtain

$$(A \hookrightarrow_Q B) \leq (A \Rightarrow_Q B)$$

which, together with (8.1) implies the equality of $\Rightarrow_Q$ and $\hookrightarrow_Q$. Assume now that there exists an $\Rightarrow_Q$ Mittelstaedt conditional in the orthocomplemented lattice $\mathcal{L}$ and let $B \leq A$ and $C \leq A^\perp$. Then

$$B = B \wedge A \leq A^\perp \vee (A \wedge B) = (A \Rightarrow_Q B)$$

furthermore

$$C \leq A^\perp \leq A^\perp \vee (A \wedge B) = (A \Rightarrow_Q B)$$

It follows that

$$B \vee C \leq (A \Rightarrow_Q B)$$

hence

$$A \wedge (B \vee C) \leq A \wedge (A \Rightarrow_Q B) \leq B$$

i.e. the Mittelstaedt conditional defined in an orthocomplemented lattice satisfies (QIC). □

A further condition which is violated by a quantum conditional is the strong transitivity:

(ST) $$(A \Rightarrow B) \wedge (B \Rightarrow C) \leq (A \Rightarrow C)$$

(If "if A implies B" and "B implies C" is true, then "A implies C" also is true.)

Proposition 8.4 *The Mittelstaedt conditional violates the* (ST) *strong transitivity condition.*

Proof: The violation of strong transitivity can be shown on a simple example. Let $\mathcal{L}_6$ be the smallest non-Boolean orthomodular lattice:

$$\mathcal{L}_6 = \{A, A^\perp, B, B^\perp, 0, I\}$$

with the partial ordering given by

$$0 \leq A \leq I, 0 \leq A^\perp \leq I, 0 \leq B \leq I, 0 \leq B^\perp \leq I$$

Then

$$\begin{aligned}(A \Rightarrow_Q I) &= I \\ (I \Rightarrow_Q B) &= B \\ (A \Rightarrow_Q B) &= A^\perp\end{aligned}$$

therefore

$$(A \Rightarrow_Q I) \wedge (I \Rightarrow_Q B) = I \wedge B = B$$

but $B \not\leq A^\perp$. □

Every conditional satisfying the law of entailment and modus ponens is weakly transitive, however, i.e. transitive not with respect to the truth but with respect to validity: the validity of $A \Rightarrow B$ and $B \Rightarrow C$ implies that $A \Rightarrow C$ is valid. Formally:

(WT)   if $(A \Rightarrow B) = I$ and $(B \Rightarrow C) = I$ then $(A \Rightarrow C) = I$

This is a straitforward consequence of the transitivity of $\leq$ and the fact that (E) and (MP) together are equivalent to the condition:

$$A \leq B \text{ if and only if } (A \Rightarrow B) = I \tag{8.3}$$

A consequence of the transitivity law (just put $A \wedge B$ in place of A) is the "law of weakening":

$$(B \Rightarrow C) \leq ((A \wedge B) \Rightarrow C)$$

The quantum conditional $\Rightarrow_Q$ does not satisfy the law of weakening either. This also can be seen on the example of the lattice $\mathcal{L}_6$, where one has:

$$(I \Rightarrow_Q B) = B \quad \text{and} \quad I \wedge (A \Rightarrow_Q B) = A^\perp$$

but $B \not\leq A^\perp$.

The typical conditionals that violate strong transitivity and weakening are the counterfactual conditionals. For instance the truth of the statement "If I did let this glas fall to the ground then the glass would break" does not entail the truth of "If I did let this glas fall to the ground and the floor were covered with a thick, soft carpet then the glass would break" is true. The violation of transitivity and weakening by $\Rightarrow_Q$ indicates that this conditional is in fact a counterfactual conditional. This is indeed the case: the quantum conditional $\Rightarrow_Q$ is a counterfactual conditional in the sense of Stalnaker's semantics. To see this we recall briefly the Stalnaker semantics of the counterfactual conditionals.

Counterfactual conditional is a $(A \Rightarrow B)$ conditional which describes the "if A were the case (were true) then B would be the case (would be true)" implication, where the proposition A stands for a condition which is not in fact true. The classical example of Lewis is "if the kangaroos had no tails then they would tople over". When would we say that this inference is true? In the worlds we know kangaroos do have tails but imagine a world differing from ours in that in this imagined world kangaroos do not have tails, which world does, however, not differ from ours very radically from the point of view of the kangaroo problem. (What this non-radical difference means, cannot be explicitly specified with absolute precision. Obviously, gravity for instance must also be part of that world, and the tail-less kangaroos, too, should be similar to those we know so that they can still be considered as kangaroos – albeit without tails). We would say that "if the kangaroos had no tails then they would tople over" is true if the tail-less kangaroos tople over indeed in that imagined world. This picture can be refined: Obviously one can imagine not one world only which differs (not too radically) from the actual one and in which kangaroos do not have tails. Assuming that all these imaginary worlds can be compared with respect to their similarity, we say that "if the kangaroos had no tails then they would tople over' is true if there is a possible (imaginary) world w in which kangaroos do not have tails and in which they tople over, and they do so in every other possible world in which they do not have tails and which is at least as similar to ours as is w.

One can formalize this idea in the following way: Let W denote the set of all possible worlds. For any $w \in W$ let a set $W(w)$ of possible worlds be given with the following properties

(i) $w \in V$ for all $V \in W(w)$
   ($W(w)$ is a centered system)
(ii) if $V_1, V_2 \in W(w)$, then either $V_1 \subseteq V_2$, or $V_2 \subseteq V_1$
   ($W(w)$ is nested system)
(iii) $W(w)$ is closed with respect to set theoretical union and intersection.

Intuitively, the $V \in W(w)$ is the set of worlds that do not differ from the world w more than to a certain degree: if $w_1, w_2 \in V$ then w_1 is at least as similar to w as is w_2.

The truth condition of the counterfactual conditional $A \Rightarrow B$ is given in this possible world semantics as follows.

Definition 8.1 The counterfactual conditional $A \Rightarrow B$ is *true at world* $w \in W$ if

(i) there exists no set of worlds V in $W(w)$ in some world of which A is true, or
(ii) there is a $V \in W(w)$ such that in a $w' \in V$ A is true, and then B also is true in every world $w' \in V$ in which A is true.

Defining $A \Rightarrow B$ to be true in case (i) is partly a matter of convention; one can perhaps defend this by saying that if A is such an absurdity that if a world in which it is true is not similar in any respect to the actual one, then in that world "everything is possible (true)".

It is customary to call the world w in which A is true an "A-world". With this convention one can give the truth condition of $A \Rightarrow B$ concisely: $A \Rightarrow B$ is true at world w if either there is no A-world similar to w, or there is an A-world w', and then B is true in every A-world which is at least as similar to w as is w'.

This possible world semantics of counterfactuals is due to D. Lewis. The Stalnaker semantics differs from this only in that the latter assumes that if there is a V in $W(w)$ that contains an A-world, then there exists a "nearest A-world", i.e. an A-world which is most similar to w. The assumption of the nearest A world is equivalent to the existence of a Stalnaker's selection function

$$S_A: W \to W$$

The interpretation of $S_A(w)$ is that it is is the nearest (to w) A-world – if there is an A world similar to w. If there is not, then $S_A(w) \equiv 0$, where the absurd world 0, where everything is true, is in W by definition. In this Stalnaker semantics the counterfactual $A \Rightarrow B$ is defined to be true at world w if and only if B is true at world $S_A(w)$.

Let us apply the Stalnaker semantics to quantum logic in the following specification: the elements of $\mathcal{P}(\mathcal{H})$ are the propositions, the possible worlds are the (vector) states (elements in $\mathcal{H}$), and let us define the Stalnaker selection function $S_A: \mathcal{H} \to \mathcal{H}$ by

$$S_A(\eta) \equiv A\eta \qquad (A \in \mathcal{P}(\mathcal{H}), \eta \in \mathcal{H}) \tag{8.4}$$

Then we have the

Proposition 8.5 *If the selection function is given by (8.4), then the Mittelstaedt quantum conditional $\Rightarrow_Q$ is a counterfactual conditional in the sense of the Stalnaker semantics.*

Proof: We must show that $A \Rightarrow_Q B$ is true in the state (possible world) η if and only if B is true in the state $S_A(\eta)$; in other words, we must show that

$$\eta \in (A \Rightarrow_Q B) \quad \text{if and only if} \quad S_A(\eta) \in B \tag{8.5}$$

Let $Ker(BA)$ be the null space of BA (the set of all vectors taken by BA to the zero element). We show first that

$$Ker(BA) = A^\perp \vee (A \wedge B^\perp) \tag{8.6}$$

If

$$\eta \in A^\perp \vee (A \wedge B^\perp)$$

then

$$\eta = \lim_i (\eta_i^{A^\perp} + \eta_i^{A \wedge B^\perp})$$

where $\eta_i^{A^\perp}$ and $\eta_i^{A \wedge B^\perp}$ are vectors from the subspaces of the corresponding projections. Since BA is bounded, it holds that

$$BA\eta = \lim_i (BA\eta_i^{A^\perp} + BA\eta_i^{A \wedge B^\perp}) = 0$$

That is
$$(A^\perp \vee (A \wedge B^\perp)) \subseteq Ker(BA)$$
Conversely, if $0 \neq \eta \in Ker(BA)$, then
$$0 = BA\eta = BA(\eta^A + \eta^{A^\perp}) = BA\eta^A$$
which implies that either $\eta^A = 0$, i.e. $\eta \in A^\perp$, or $\eta^A \in B^\perp$, and then $\eta^A \in (A \wedge B^\perp)$. All this implies
$$Ker(BA) \subseteq A^\perp \vee (A \wedge B^\perp)$$
Let $A\eta \in B$. This holds if and only if
$$BA\eta = A\eta$$
which holds if and only if
$$(B - I)A\eta = 0$$
which holds if and only if
$$B^\perp A\eta = 0$$
which holds if and only if
$$\eta \in Ker(B^\perp A)$$
which by (8.6) holds if and only if
$$\eta \in (A^\perp \vee (A \wedge B))$$
□

Note that the selection function $S_A(\eta) = A\eta$ chosing the nearest (to η) A-world picks for η the vector in the range of A which is nearest in Hilbert space norm to η. The measure of similarity of the possible worlds is thus the metric (norm) in $\mathcal{H}$.

8.2. Conditional probability and statistical inference

The problem of statistical inference, informally formulated, is the problem of what probabilities should be assigned to certain events on the basis of probabilites of certain other events. The classical case is when the probabilities $\mu(A_i)$ of certain events are given as background information, and one "learns" that an event B has probability equal to 1. The question is how to revise the probabilities $\mu(A_i)$ on the basis of this new information.

A possible answer to this question is: let the new probabilities $\nu(A_i)$ be the conditional probabilities of A_i with respect to B, given by the Bayes' rule:

"Bayes' rule": $$\nu(A_i) \equiv \frac{\mu(A_i \cap B)}{\mu(B)} \qquad (8.7)$$

More generally, but still remaining within the framework of classical (i.e. commutative) probability theory, the problem of statistical inference can be formulated in the following way.

Let (Γ, Ω, μ) be a classical probability space, i.e. Ω a σ-algebra of subsets of Γ and μ a probability measure on Ω. Let Ω_0 be a sub-σ-algebra of Ω, and let us assume that a probability measure ν_0 is given on Ω_0. The probability measure μ is interpreted as representing the background information, and the probabilities $\nu_0(A_0)$ of events A_0 belonging to Ω_0 are considered as given or obtained (e.g. by measurements) as new information. The question of statistical inference is how to extend ν_0 from Ω_0 to the whole Ω. The classical answer to this question based on the Bayes' rule is the theory of classical conditionalization:

Let $L^\infty(\Gamma, \Omega, \mu)$ and $L^\infty(\Gamma, \Omega_0, \mu_0)$ (μ_0 being the restriction of μ to Ω_0) be the commutative von Neumann algebras of Ω resp. Ω_0 measurable essentially bounded functions defined on Γ. Recall that $f \in L^\infty(\Gamma, \Omega, \mu)$ if there is a constant c such that $|f(x)| \leq c$ for μ-almost all $x \in \Gamma$. Recall also that $L^\infty(\Gamma, \Omega, \mu)$ is a Banach space with the norm

$$\|f\|_\infty = \inf\{c \mid |f(x)| \leq c \; \mu\text{-almost everywhere}\}$$

If $L^1(\Gamma, \mu)$ denotes the integrable functions on Γ, then $L^\infty(\Gamma, \Omega, \mu)$ is the Banach dual space of $L^1(\Gamma, \mu)$. The isomorphism

$$h: L^\infty(\Gamma, \Omega, \mu) \to (L^1(\Gamma, \mu))^*$$

is given by

$$(h(f))(g) = \int fg \, d\mu \qquad g \in L^1(\Gamma, \mu)) \qquad (8.8)$$

Fixing an $f \in L^\infty(\Gamma, \Omega, \mu)$ one can define a μ_f functional on $L^1(\Gamma, \mu_0)$ by using the isomorphism h given by (8.8):

$$\mu_f(g_0) \equiv \int f g_0 \, d\mu$$

and by the duality
$$L^1(\Gamma, \mu_0))^* = L^\infty(\Gamma, \Omega_0, \mu_0)$$
there exists an $f_0 \in L^\infty(\Gamma, \Omega_0, \mu_0)$ such that

$$\mu_f(g_0) \int f g_0 \, d\mu = \int f_0 g_0 \, d\mu_0 \qquad g_0 \in L^1(\Gamma, \mu_0)$$

The map
$$f \mapsto T(f) \equiv f_0$$
is a positive, linear, unit preserving projection from $L^\infty(\Gamma, \Omega, \mu)$ onto $L^\infty(\Gamma, \Omega_0, \mu)$, and it has the property
$$\int f \mathrm{d}\mu = \int Tf \mathrm{d}\mu$$

If Ω_0 is generated by a countable set $\{A_i \in \Omega\}$ of pairwise disjoint elements A_i, then
$$T(\chi_B) = \sum_i \frac{\mu(B \cap A_i)}{\mu(A_i)} \chi_{A_i} \qquad B \in \Omega \tag{8.9}$$

That is, in this special case T can be given explicitly on the characteristic functions χ_B by the "Bayes's rule" (8.7); however, for an arbitrary Ω_0 the map T cannot be given explicitly (the existence of T in the general case is a consequence of the Radon-Nikodym existence theorem, which is the essential ingredient in proving the duality $(L^1(\Gamma, \mu_0))^* = L^\infty(\Gamma, \Omega_0, \mu_0)$).

The map T is called the (Ω_0, μ)-*conditional expectation* from $L^\infty(\Gamma, \Omega, \mu)$ onto $L^\infty(\Gamma, \Omega_0, \mu_0)$, and T is the generalization in classical commutative probability theory of the concept of conditional probability as defined by the Bayes' rule. Further terminology:

- $T(f)$ is the (Ω_0, μ)-*conditional expectation* of f;
- $T(\chi_A)$ is the (Ω_0, μ)-*conditional probability* of $A \in \Omega$.

This latter $T(\chi_A)$ is not a number but a function. One obtains "real", i.e. numerical conditional probabilities by applying a probability measure to $T(\chi_A)$. This leads then to the general answer to the question on statistical inference in commutative probability: Given ν_0 on Ω_0 let the extension ν of ν_0 be defined by

$$\nu(A) \equiv \nu_0(T(\chi_A)) \qquad (A \in \Omega)$$

where T is the (Ω_0, μ)-conditional expectation from the function space $L^\infty(\Gamma, \Omega, \mu)$ to $L^\infty(\Gamma, \Omega_0, \mu_0)$.

This prescription of conditionalization can be summed up in the terminology of operator algebras in the following way: Let $\mathcal{M}$ be a *commutative* von Neumann algebra, φ be a state on $\mathcal{M}$, and let $\mathcal{M}_0$ be a von Neumann subalgebra of $\mathcal{M}$. Assume that a state ψ_0 is given on $\mathcal{M}_0$. The problem of statistical inference is the problem of how to extend ψ_0 to $\mathcal{M}$ with respect to a state φ, which is to be interpreted as the background information. Answer: put

$$\psi(A) \equiv \psi_0(TA) \qquad (A \in \mathcal{M})$$

where $T: \mathcal{M} \mapsto \mathcal{M}_0$ is a φ-preserving conditional expectation, i.e. T is a positive, linear, unit preserving projection from $\mathcal{M}$ onto $\mathcal{M}_0$ such that $\varphi(T(A)) = \varphi(A)$. If $\mathcal{M}$ is commutative, then such a T always exists, since a commutative von Neumann algebra is isomorphic to an $L^\infty(\Gamma, \Omega, \mu)$, and then one can conditionalize in the way described.

This formulation (but not the solution!) of the problem of statistical inference remains meaningful if we drop the assumption of commutativity: Let $\mathcal{M}$ be an arbitrary von Neumann algebra, φ be a state on $\mathcal{M}$, and let $\mathcal{M}_0$ be a subalgebra of $\mathcal{M}$ with common unit. Assume that a state ψ_0 is given on $\mathcal{M}_0$. What should be the extension of ψ_0 to $\mathcal{M}$?

On the basis of the commutative case one would want to answer: let the extension ψ be given by $\psi(A) = \psi_0(TA)$ with a φ preserving projection from $\mathcal{M}$ onto $\mathcal{M}_0$. Unfortunately, such a T does not exist in general in the non-commutative case. For instance if $\mathcal{M} = M_2 \otimes M_2$ (M_2 being the algebra of complex two-by-two matrices) and φ is a state on $\mathcal{M}$, then a φ preserving projection from $\mathcal{M}$ onto $M_2 \otimes I$ exists only if φ is a product state:

$$\varphi(AB) = \varphi_1(A)\varphi_2(B) \quad A \in M_2 \otimes I, B \in I \otimes M_2$$

for some states φ_1 and φ_2 on the component algebras.

The fact that a φ-preserving conditional expectation from a von Neumann algebra $\mathcal{N}$ to a von Neumann subalgebra $\mathcal{N}_0$ does not exist in general in the non-commutative case poses a problem for what is called "Bayesianism" in that it blocks a straightforward extension of Bayesianism to non-commutative event structures. Bayesianism (in its commutative version) is the doctrine characterized by the following three assumptions.

1. The probability measures μ, which are supposed to be defined on a Boolean algebra Ω of events, are interpreted as the measures of rational belief of an abstract rational person (usually and hereafter called "agent") who is supposed to be capable of ideally logical thinking;
2. The changes in the agent's degree of belief, $\mu \to \mu'$, on his learning an event E in Ω to be true (i.e having probability one) are given by conditionalization on E via the Bayes' rule (8.7):

$$\mu'(X) = \mu(XE)/\mu(E) \equiv \mu(X/E) \quad \text{for all } X \text{ in } \Omega$$

or, more generally, if the agent happens to learn the probabilities of events of a whole subfield Ω_0 of Ω rather than the probability of a single event E, then

$$\mu'(X) = \mu(TX) \quad X \in \Omega$$

where T is the (Ω_0, ν)-conditional expectation from the commutative von Neumann algebra $L^\infty(\Omega, \nu)$ onto its von Neumann subalgebra

$L^\infty(\Omega_0, \nu_0)$ with respect to an "a priori" probability measure ν on Ω (and its restriction ν_0 to Ω_0). The measure ν is to be interpreted as the representative of the agent's background information, and it is preserved under T.

Note the following *stability property* of Bayesian statistical inference: If the agent reviews his *new* degrees of belief (μ') in the light of the *same* evidence E *again*, he must conclude that

$$\mu''(X) = \mu'(XE)/\mu'(E) = [\mu(XEE)/\mu(E)]/[\mu(XE)/\mu(E)] = \mu'(X)$$

that is to say, the agent's newly revised degrees of belief (μ'') do not differ from what he had already inferred from μ on the basis of learning E. In the general case this stability is expressed and ensured by the fact that T is a ν-preserving *projection* from $L^\infty(\Omega, \nu)$ onto $L^\infty(\Omega_0, \nu_0)$:

$$TTf = Tf \qquad \text{for all} \qquad f \text{ in } L^\infty(\Omega, \nu)$$

(The events, i.e. the elements of Ω are identified with the characteristic functions in $L^\infty(\Omega, \nu)$).

This stability seems to be essential in Bayesian statistical inference, for let us assume that it does not hold. In this case the agent would find himself in a very frustrating situation every time after having learned the probabilities of the events in Ω_0, since he would have his new degrees of belief being μ' after the first inference, but, looking at μ' at a second time, and without having gained any new evidence, he would have to conclude that his new degrees of rational beliefs should be rather $\mu'' \neq \mu'$ than μ'. Given ν, μ and S_0 as the exclusive basis for the inference, the agent could not decide rationally between μ' and μ'', i.e. either he should choose one of them irrationally (by tossing a coin for instance), in which case the chosen new degrees of belief could no longer be considered as degrees of *rational* belief, or the agent's degrees of belief become indefinite. In either case, obviously, the inference would no longer be Bayesian in the sense of the above definition. In short: if the probability measures are interpreted as measures of degree of rational belief, then the stochastic inference $\mu \to \mu'$ must be stable.

Since a φ-preserving conditional expectation from a Neumann algebra $\mathcal{N}$ to a von Neumann subalgebra $\mathcal{N}_0$ does not exist in general for arbitrary φ and $\mathcal{N}_0$ in the non-commutative case, Bayesianism cannot be extended to non-commutative event structures in a straightforward manner. If Bayesianism cannot be extended at all to the non-commutative case, then a possible conclusion is that probability measures defined on a non-commutative event structure cannot be interpreted as degrees of rational belief.

This conclusion would entail that the probabilities occurring in quantum mechanics cannot be reasonably considered as degrees of rational belief, or, identifying the numbers $R(E) = 1 - p(E)$ with the measure of lack of knowledge concerning the quantum event E (where $p(E)$ is the degree of rational belief), one could equally say that quantum probabilities cannot be reasonably viewed as measures of ignorance. This would be a further argument against the reasonability of what is called the "ignorance interpretation" of quantum state, and which has already been criticized on quite different grounds [105], [118].

The next proposition characterizes an important special case when a state preserving conditional expectation exists.

Proposition 8.6 *Let $\mathcal{M}_0$ be a von Neumann subalgebra (with common unit) of the von Neumann algebra $\mathcal{M}$, and assume that there exists a faithful, normal, tracial state τ on $\mathcal{M}$. Then there exists a $T: \mathcal{M} \to \mathcal{M}_0$ linear map having the following properties:*

(i) $T(I) = I$
(ii) *if $X \geq 0$ then $T(X) \geq 0$*
(iii) $\tau(T(X)) = \tau(X)$ *for all $X \in \mathcal{M}$*
(iv) $T(XA_0) = T(X)X_0$ *for $X \in \mathcal{M}$, $X_0 \in \mathcal{M}_0$*
(v) $T(X_0) = X_0$ *for all $X_0 \in \mathcal{M}_0$*

The map T is called the τ-preserving conditional expectation (or projection of norm one) from $\mathcal{M}$ onto $\mathcal{M}_0$

Proof: We give T explicitly in the GNS representation $(\mathcal{H}_\tau, \Phi_\tau, \pi_\tau)$ determined by the faithful tracial state τ (this representation is also called the *standard representation* of $\mathcal{M}$). Recall that the Hilbert space $\mathcal{H}_\tau$ is the completion of $\mathcal{M}$ with respect to the inner product $\langle X, Y \rangle = \tau(X^*Y)$, Φ_τ denotes the vector in $\mathcal{H}_\tau$ corresponding to the unit in $\mathcal{N}$, and the representation π_τ of $\mathcal{N}$ on $\mathcal{H}_\tau$ is defined by

$$\pi_\tau(X)Y = XY \qquad X, Y \in \mathcal{A}$$

(see the Section 6.1 in Chapter 6). $\mathcal{M}_0$ is then a closed linear subspace in $\mathcal{H}_\tau$, thus there is an orthogonal projection P to this subspace. Let

$$T: \pi_\tau(\mathcal{M}) \to \pi_\tau(\mathcal{M}_0)$$

be defined by

$$T(\pi_\tau(X))\xi = P\pi_\tau(X)\xi \qquad (\xi \in \mathcal{H}_\tau) \tag{8.10}$$

The projection P is the identity operator in $\mathcal{M}_0$, so it commutes with every operator in $\pi_\tau(\mathcal{M}_0)$. It follows that for any $Z \in \pi_\tau(\mathcal{M}_0)'$ the three elements $P, \pi_\tau(X)$ and Z commute, thus $P\pi_\tau(X)$ commutes with Z i.e.

$$(P\pi_\tau(X)) \in \pi_\tau(\mathcal{M}_0)'' = \pi_\tau(\mathcal{M}_0)$$

The properties (i),(iii),(iv) and (v) hold obviously for this T. To see the positivity condition (ii) one has to see that if for every $X \in \mathcal{M}$ one has

$$\langle \pi_\tau(X)\Phi_\tau, \pi_\tau(A)\pi_\tau(X)\Phi_\tau \rangle \geq 0 \tag{8.11}$$

then for every $Y \in \mathcal{M}$ one has

$$\langle \pi_\tau(Y)\Phi_\tau, P\pi_\tau(A)\pi_\tau(Y)\Phi_\tau \rangle \geq 0 \tag{8.12}$$

(8.11) is just $\tau(X^*AX) \geq 0$ and (8.12) is $\tau(Y^*PAY) \geq 0$. Using the trace property of τ we have

$$0 \leq \tau(X^*AX) = \tau(XX^*A) \qquad X \in \mathcal{M} \tag{8.13}$$

and

$$\tau(Y^*PAY) \tag{8.14}$$
$$= \tau(YY^*PA) \ = \ \tau(YY^*PAP) \tag{8.15}$$
$$= \tau(PYY^*PA) \ = \ \tau(PY(PY)^*A) \tag{8.16}$$

and so (8.13) implies $\tau(PY(PY)^*A) \geq 0$. $\square$

As the proof of the previous proposition shows, the tracial property of τ is used in showing that the T defined by (8.10) is a positie map. Replacing the tracial state τ by a general state ϕ, the formula (8.10) could be used to define a map in the GNS representation determined by ϕ, that map would not be positive, however. For a general $\mathcal{M}, \mathcal{M}_0$ and ϕ, a ϕ-preserving projection from $\mathcal{M}$ onto $\mathcal{M}_0$ does not exist. If, however, one drops the requirement that $T: \mathcal{M} \to \mathcal{M}_0$ preserve a state, then a T projection might exist.

For instance let $\mathcal{M} = \mathcal{B}(\mathcal{H})$ and $Q = \sum_i q_i E_i$ be an observable with discrete spectrum and spectral projections E_i. If $\mathcal{M}_0$ is the subalgebra given by

$$\mathcal{M}_0 \equiv \{A \in \mathcal{B}(\mathcal{H}) \mid AE_i = E_i A \ (i = 1, 2 \ldots)\} \tag{8.17}$$

One then can define T by

$$T(A) \equiv \sum_i E_i A E_i \tag{8.18}$$

This T is known as the projection postulate. T is a positive, linear, unit preserving projection from $\mathcal{B}(\mathcal{H})$ onto $\mathcal{M}_0$, it corresponds to the classical conditional expectation. In analogy with the classical solution of the statistical inference problem it is natural to extend ψ_0 from $\mathcal{M}_0$ by $\psi(A) \equiv \psi_0(TA)$, where T is the conditionl expectation defined by (8.18).

In particular, if $\mathcal{M}_0$ is generated by a single projection E, and $A \in \mathcal{P}(\mathcal{H})$ is a proposition, then

$$T^E(A) = EAE + E^\perp A E^\perp$$

thus if ψ is a state, then

$$\psi(A) = \psi(EAE + E^\perp A E^\perp)$$

can be interpreted as the non-commutative $\mathcal{M}_0$-conditionalization of the probability of A.

8.3. Breakdown of Stalnaker's Thesis in quantum logic

The statistical inference "If ψ_0 then ψ", $\psi_0 \to \psi$, connects the dual spaces (state spaces) of $\mathcal{M}_0$ and $\mathcal{B}(\mathcal{H})$. The question arises whether this $\to$ arrow between the dual spaces is the dual of an $\Rightarrow$ arrow (conditional) defined in $\mathcal{P}(\mathcal{H})$. In the simplest case this question is the question whether

$$\psi(A \Rightarrow E) = \psi(T^E A) \tag{8.19}$$

holds, i.e. whether it is the case that

"probability of conditional= conditional probability"

The claim that "probability of conditional= conditional probability" is called in the philosophy of science literature the "Stalnaker Thesis". The thesis is plausible. What else could be the "probability of throwing six with a dice if the throw is an even number" if not "the probability of 'if I throw an even number then it will be six'". Plausible or not, once $\Rightarrow$ and T^E are formally determined, the validity or non-validity of the thesis is also fixed. It is natural to investigate the question of validity of the thesis in the case when $\Rightarrow$ in (8.19) is replaced by the quantum conditional.

Call it "Strong Stalnaker's Thesis" the claim that (8.19) holds for every ψ and A, E, "weak Stalnaker Thesis" the claim that there is a ψ for which (8.19) holds for every A and E.

Proposition 8.7 *If $\Rightarrow$ is the Mittelstaedt conditional, then neither the strong nor the weak Stalnaker Thesis holds.*

Proof: Consider

$$\varphi(A^\perp \vee (A \wedge B)) = \varphi(ABA + A^\perp B A^\perp) \tag{8.20}$$

This cannot hold for every state φ and for all A, B, since, if $A \leq B$, then $A \wedge B = A$, and the projection on the left hand side of (8.20) is the identity, whereas the right hand side is B. If $B \neq I$, then the projection on the left and right hand sides of (8.20) differ, and so the left and right hand sides of (8.20) differ for every fathful state φ for $B \neq I$. (The state φ is faithful if $\varphi(A) > 0$ for every $0 < A$) Let now φ be an arbitrary state, and let $s(\varphi)$ be the support projection of φ (i.e. $I - s(\varphi) = s(\varphi)^\perp$ is the least upper bound of projections E for which $\varphi(E) = 0$). Since φ is non-zero $s(\varphi) > 0$, thus there is a non-zero $E \in \mathcal{P}(\mathcal{H})$, such that $E \leq s(\varphi)$. Substituting $A = s(\varphi)^\perp$ and $B = E^\perp$ into (8.20) one obtains

$$\varphi(s(\varphi) \vee (s(\varphi)^\perp \wedge E^\perp)) = \varphi(s(\varphi)E^\perp s(\varphi) + s(\varphi)^\perp E^\perp s(\varphi)^\perp) \qquad (8.21)$$

Since $E \leq s(\varphi)$, it holds that $E^\perp \geq s(\varphi)^\perp$, thus

$$s(\varphi)^\perp \wedge E^\perp = s(\varphi)^\perp$$

Hence the left hand side of (8.21) is equal to 1, whereas the right hand side is $\varphi(E^\perp)$, i.e. $\varphi(E) = 0$. On the other hand, making the substitution $A = s(\varphi)$ and $B = E$ one obtains in a similar manner $\varphi(E) = 1$. □

One may ask: Is the breakdown of Stalnaker's Thesis perhaps due to the fact that the definition of quantum conditional probability by T^E is inappropriate? This question seems to be the more justified since T^E is suspect: It is known that the projection postulate (i.e. the definition (8.18) fails to remain valid if the operator Q does not have pure discrete spectrum. If this is the case, then (8.18) becomes not only meaningless but it cannot even be made meaningful in the sense that if one defines the subalgebra

$$\mathcal{M}_0 \equiv \{A \in \mathcal{B}(\mathcal{H}) \mid A\mathbf{P}^Q(d) = \mathbf{P}^Q(d)A, \ d \in \mathcal{B}(\mathbb{R})\}$$

in complete analogy with (8.17), and Q does not have a pure point spectrum, then there does not exist a positive, linear unit preserving *projection* from $\mathcal{B}(\mathcal{H})$ onto $\mathcal{M}_0$. The question is what should replace T in this case. This question, the problem of conditionalization in non-commutative probability theory is a deep question of the non-commutative probability. Formally put the question is what property (or properties) of the T projection should we give up so that $\psi_0 \circ T$ still be a state. Clearly, since states are *positive* and *linear* functionals, a minimal requirement is that T remain a positive, *linear* map, hence the only property that can be sacrificed is that T is a projection. If, however T is a positive, linear map, then the Stalnaker's Thesis does not hold, as we show next:

Proposition 8.8 *If the conditional expectation T^E is replaced by a linear map T, then the non-commutative Stalnaker's Thesis (8.19) is violated in any state φ if $\Rightarrow$ is the Mittelstaedt conditional.*

Proof: If (8.19) did hold for a state φ for every projection, then

$$\begin{aligned}\varphi(A^\perp \vee (A \wedge (F+G))) &= \varphi(T(F+G)) \\ = \varphi(A^\perp \vee (A \wedge F)) &+ \varphi(A^\perp \vee (A \wedge G))\end{aligned} \qquad (8.22)$$

holds for all projections $A, F, G \in \mathcal{P}(\mathcal{H})$ for which both sides of (8.22) is meaningful, i.e. for which $(F+G) \in \mathcal{P}(\mathcal{H})$. Assume that φ is a faithful state. Let A be an arbitrary non-trivial projection and consider the projections $F = A$ and $G = A^\perp$. Then

$$\begin{aligned} A^\perp \vee (A \wedge F) + A^\perp \vee (A \wedge G) &= A^\perp \vee (A \wedge A) + A^\perp \vee (A \wedge A^\perp) \\ = I + A^\perp \neq A^\perp \vee (A \wedge (F+G)) &= A^\perp \vee (A \wedge (A + A^\perp)) = I \end{aligned}$$

Since φ is a faithful state $\varphi(A^\perp) > 0$, and so (8.22) is violated. Assume that φ is not faithful. Then $0 < s(\varphi) < I$. Let F be an arbitrary non-trivial projection. If (8.22) holds for every projection for which the left hand side is meaningful, then we obtain

$$\begin{aligned} 2 &= \varphi(s(\varphi)) + \varphi(s(\varphi)) \\ &\leq \varphi(s(\varphi) \vee (s(\varphi)^\perp \wedge F)) + \varphi(s(\varphi) \vee (s(\varphi)^\perp \wedge F^\perp)) \\ &= \varphi(s(\varphi) \vee (s(\varphi)^\perp \wedge (F + F^\perp))) = \varphi(I) = 1 \end{aligned}$$

and this is not possible. $\square$

The limits of breakdown of the Stalnaker's Thesis can be approached from the other direction as well: One can ask, how restrictive the validity of (8.19) is for the conditional operation. In other words, one may ask whether the Thesis can be saved by putting the conditionals $\Rightarrow_3$ or $\Rightarrow_2$ in place of the Mittelstaedt conditional. The answer to this question is also negative:

Proposition 8.9 *The non-commutative Stalnaker's Thesis* (8.19) *does not hold for every A, B, if $\Rightarrow$ is either of $\Rightarrow_3$ and $\Rightarrow_2$.*

Proof: Let φ be an arbitrary state. Then there exists a $B \geq s(\varphi) \neq 0$ projection. Take a projection A orthogonal to B. Then the left hand side of (8.19) is $\varphi(B)$ either with $\Rightarrow_3$ or with $\Rightarrow_2$ in place of $\Rightarrow$, whereas the right hand side of (8.19) is equal to zero. $\square$

The Proposition 8.9 tells us that the Stalnaker's Thesis does not hold even in the weak sense if the conditional probabilities are given by conditional expectations and the conditional is one that satisfies minimal implicative criteria. Thus the non-commutative (quantum) statistical inference cannot be related to the non-commutative (quantum) conditional in the manner of Stalnaker's Thesis. It is not known, in what way the "logic" of statistical inference over a non-commutative event structure can be related to the logic of the underlying logical structure.

8.4. Bibliographic notes

Section 8.1 is based on the reviews [60] and [61], which, however, do not contain the proofs of all propositions. Proposition 8.3 is due to Mittestaedt [102]. Further characterization of the $\Rightarrow_Q$ quantum conditional can be found in [62]. The classic reference for the possible world semantics of counterfactuals is [92]. The Stalnaker's Thesis and papers on its interpretation are collected in [64]. The Thesis was investigated from the point of view of non-commutative logic/probability by Rehder in [135]. The Propositions 8.7, 8.7 and 8.9 are taken from [125]. For the theory of classical conditionalization see [94], for a review of the non-commutative conditional expectation see [109]. That there exists no projection from $\mathcal{B}(\mathcal{H})$ onto von Neumann subalgebra generated by the spectral projection of an operator with continuous spectrum is shown in [44].

CHAPTER 9

The problem of hidden variables

9.1. Historical remarks

The problem of hidden variables in quantum mechanics is almost as old as quantum mechanics itself. The first mention of the idea goes back to Born's 1926 paper [26] that introduces the probabilistic interpretation of quantum state (see also [73]), and the hidden variable problem has revived from time to time in the past six decades. We could witness its last revival in the early eighties, and the debates about this issue have been going on ever since.

One may find it strange that the problem has not been settled yet despite its long history and the enormous literature dealing with it. This openness of the problem is explained by the fact that this problem is not a "purely" physical one; it is partly a philosophical problem.

One sign of the non-purely physical character of the problem is that the problem itself cannot even be precisely formulated once and for all, not even in one and the same framework of quantum mechanics. In addition, the problem arises taking on new forms with the conceptual development of quantum mechanics, and it should be (and has indeed been) re-phrased in every new framework that have been created.

The hidden variable problem has commonly been associated with the question of validity of certain very general philosophical (metaphysical) principles, such as the principle of determinism or the philosophical standpoint known as realism. In view of the seriousness of very far reaching claims that have been made with reference to the hidden variable problem, it has remained surprisingly unclear in the literature, what exactly the link is between the hidden variable problem and the philosophical doctrines in question. Typically, one interprets results asserting the *nonexistence* of a T' hidden theory having certain features as a scientific evidence for (or even "proof" of) *invalidity* of some philosophical doctrines, and it also is quite common to interpret results spelling out the *existence* of some kind of a hidden theory T' as a scientific support in favor of the philosophical principle in question.

One can see this on the example of what has become known as "von Neumann's impossibility proof", which is the first definite result in this field, and which has become a classic subject of controversy in connection with the hidden variable problem: von Neumann's result has often been interpreted as a scientific theorem falsifying determinism (von Neumann himself made statements in his book in this spirit), and the existence of simple "hidden variable models" constructed deliberately as counterexamples for von Neumann's impossibility proof have been viewed by many as results showing that nothing is wrong with the principle of determism, and that something is wrong with von Neumann's proof. Surprizingly, one does not find in the literature a detailed analysis and reconstruction of von Neumann's impossibility proof, however. Attempting to do this here in detail would lead us too far away from the central theme of this work, so we are content with making only a couple of non-technical remarks on von Neumann's treatment of the problem. Somewhat more will be said on von Neumann's result in Section 9.2.

The key observation necessary to understand the significance of von Neumann's impossibility proof is what has already been emphasized e.g. by Bub [30], namely that von Neumann investigates the statistical character of quantum mechanics and the problem of quantum mechanical probabilities *in connection with the probabilities occurring in classical statistical mechanics.* (This context is explicitely stated by von Neumann, too, in the preface to his book.) The significance of keeping in mind this context of von Neumann's proof comes from the fact that the concept of state in classical statistical mechanics already raises the problem of determinism in that it contradicts the state concept in classical mechanics: the classical statistical mechanical state is (in today's terminology) a probability measure. The reason why this probability measure should indeed be viewed as a physical state is that it can be identified with thermodynamical states – as far as one can derive thermodynamic relations with its help – on the other hand, this state concept is incompatible with the state concept of classial mechanics because, unlike in the case of classical mechanical state (which is identified with a single point in the phase space), the physical quantities possess a non-zero dispersion in a statistical mechanical state. In this situation a physicist (or a philosopher) has two options:

1. Insisting on accepting classical statistical mechanical states as physical states;
2. Rejecting that classical statistical mechanical states are genuine physical states.

Clearly, while 1. entails that one has to modify (or abandon) the principle of determinism understood along the conceptual lines rooting in classical mechanics; the option 2. does not force one even to rethink, much less to sacrifice determinism.

It is not surprising that it was this second option that the overwhelming majority of physicists had chosen, and this is the view which is still widely accepted. The way physicists reject classical statistical mechanical states as real physical states is by accepting the following well-known reasoning: A classical statistical mechanical system consists of a large number of particles interacting and moving according to the laws of classical mechanics and, therefore, the point representing the whole system in the phase space also moves deterministically. To describe the motion of the phase point would require both the exact knowledge of the initial states of all the particles and the ability to solve a large number of differential equations. But one is unable to solve so many equations of motions, and the initial conditions are not known either. *For this reason*, as the argument goes, we describe the system by probabilities, although the real physical system is at every moment in some well defined state as understood in classical mechanics. (This view is expressed in [5] for instance.)

Now, these well defined real physical states in the sense of classical mechanics, i.e the single phase points can be identified with the Dirac measures concentrated at these points, and the Dirac measures are but the *pure and dispersion-free* states in classical statistical mechanics. Consequently, the classical statistical mechanical states are considered in this interpretation as signs of our inability to give the precise real classical statistical mechanical (pure) state the system is in, the probabilities are viewed as measures of our ignorance concerning these "real" states.

By putting forward this ignorance interpretation of classical statistical mechanical probabilities one hopes to have achieved two goals: Both the appearance of probabilites seems to be explained, and the clash between the state concept of classical statistical mechanics and determinism (understood as the state concept of classical mechanics) seems resolved.

It is quite understandable that, after the statistical character of quantum mechanics had become clear, one tried to do away with it in the same manner as one did in the case of classical statistical mechanics, i.e. by saying that the "real" physical quantum states are the pure quantum states, and the probabilities are but the measures of our ignorance. This idea was not alien to von Neumann either, for he had become familiar with the problems of classical statistical mechanics. Judged on the basis of the dates of the appearance of his published works, von Neumann was working on problems of classical statistical mechanics just during the preparation of his book on the mathematical foundations of quantum mechanics: the three decisive papers his book is based on appeared in 1927, and there is no mention of the hidden variable problem and of the impossibility proof in any of them. Between 1927 and 1932, the year of publication of the book, von Neumann published his proof of the ergodic theorem and H-theorem in

1929, the proof of the quasi ergodic hypothesis in 1932, and three further papers on problems arising in classical statistical mechanics.

The historic significance of von Neumann's impossibility proof is that he had proven that the ignorance interpretation does not work in quantum mechanics, it does not work in the same way as in classical statistical mechanics anyway; that is to say, though one may formulate an ignorance interpretation of quantum state, this does not help one out of the crunch: pure quantum states are not dispersion-free, thus one has to meet the challenge of interpreting probability in quantum mechanics. What von Neumann did *not* prove is that there is no probability interpretation, and, consequently, no conceivable interpretation of quantum state, with the help of which the classical concept of determinism can be reconciled with quantum mechanics – simply because such a general statement cannot be proven in the strict sense of "proof".

The history of the hidden variable problem after von Neumann is full of debates and, sometimes, of misunderstandings as well. Some of those feeling the need of a redefinition of the hidden variable problem criticize von Neumann's approach and try to show the irrelevance of von Neumann's impossibility proof by constructing hidden variable theories [22], [24], [10] [11]. In the majority of these critiques the hypothetical T' hidden theory of quantum mechanics is considered a physical theory which may differ completely and radically from quantum mechanics, and the charge is raised against von Neumann that he assumed the validity of quantum mechanical rules in T', which von Neumann did not try to justify, and which is not, in fact, justifiable. In particular, von Neumann's assumption that the hidden states are linear functionals on not compatible observables was considered in the above mentioned works as lacking any support. As a result, von Neumann's statement was viewed in these critiques as having nothing to do with the existence of a T' hidden theory so envisioned. Furthermore, the concept of hidden theory T' that is possibly different entirely from quantum mechanics was entangled with the more or less tacit assumption of the existence of a sub-quantum physical reality, which T' was supposed to be a deterministic description of.

As a reaction to the critiques of von Neumann's impossibility proof there appeared attempts aiming at producing stronger impossibility proofs by weakening von Neumann's assumptions that had been questioned. These new approaches were made possible by the fact that quantum mechanics did not stop developing after 1932, and by the sixties there existed two, non-equivalent axiomatic approaches to quantum mechanics: the one based on lattice theory and the other one formulated in terms of the operator algebra theory. It is the theory of von Neumann algebras that connects these two approaches.

In the lattice theoretic framework von Neumann's proof was generalized by [74]. It is clear, however, that if T' is considered to be a theory different from quantum mechanics, then any impossibility proof that remains within a given, well defined framework of quantum mechanics, such as Jauch and Piron's, can be subjected to the a same critique that was directed against von Neuamnn's proof. Not surprisingly then, Jauch and Piron's treatment of the problem was soon attacked by Bohm and Bub [25]. That the arguing parties have different views regarding what a hidden theory of quantum mechanics is, or is not supposed to be, was emphasized by Gudder [54], [55].

Gudder's attitude towards the hidden variable problem should be considered a turning point in the history of the hidden variable issue: he deliberated the problem from the burden of speculation over the existence or nonexistence of some sub-quantum reality and its description by T', and he approached the problem with a different philosophy. Gudder argues that one should not ask whether there are hidden parameters as some kind of really existing physical quantities or properties, and, according to Gudder, Bohm and Bub are right in saying that the existence of such entities can never be ruled out with absolute and final certainty; rather, argues Gudder, the question should be asked as to whether a certain model or description of the physical reality (and in what sense) admitts hidden parameters. Thus any statement regarding the hidden variables is, in this interpretation, a meta-theorem, subject of wich is not the reality but a given description of it. Consequently, it may very well happen that certain models admit hidden parameters, others do not, and it should not surprise anyone if it turns out that one and the same model admits hidden variables and, at the same time, it does not - with respect to different definitions of hidden variables of course. Gudder draws the attention to the definition-sensitiveness of the hidden variable problem by proving, in the lattice theoretic framework, both positive and negative results on the existence of hidden variables. The approach adopted in the present chapter follows Gudder's philosophy.

Neumann's definition of hidden variables (and Neumann's imposssibility proof) was generalized in the operator algebraic approach by Misra [101]. Misra's results will be discussed below in detail.

Bell's work is considered in the literature as another turning point in the history of the hidden variable problem for at least two reasons. One being that Bell connected the hidden variable problem with the issue of physical locality, and the other being that his way of treating the problem of compatibility of the asssumption of hidden variables and physical locality made it possible to verify empirically the consequences of assuming a local hidden variable theory of quantum mechanics. Bell's contribution will be recalled briefly in Section 9.4.

9.2. Notion of and no-go results on dispersive hidden theories

9.2.1. DEFINITION OF DISPERSIVE HIDDEN THEORY

The problem of "hidden variables", or, more precisely, the problem of the theoretical possibility of interpreting quantum mechanics through assuming some kind of hidden parameters, was originally motivated by the statistical character of quantum mechanical state description. The question was formulated whether it was possible, in principle at least, to construe a T' theory "behind" the T theory of quantum mechanics that can eliminate the statistical character of quantum states and which, at the same time, is capable of reproducing the quantum mechanical predictions. The following "Definition" summarizes the intuitive requirements that a T' hidden theory for a theory T should fulfill.

Definition 9.1 Let T and T' be two physical theories describing the physical system S in a statistical manner by state variables $v \in E(T)$ and $v \in E(T')$ in T and T' respectively. The state variables v' are said to be "hidden (state) variables" for T, and T' is called a *hidden theory* of T if the following 1.-3. conditions are fulfilled

1. within the framework of T, the variables v' are not considered accessible through direct observation;
2. each v in $E(T)$ can be obtained through some kind of an averaging over variables in $E(T')$ in such a way that
3. if σ_v and $\sigma_{v'}$ are some measures of how uncertain the state description of S by the respective variables v and v' are, then $\sigma_{v'}$ is smaller than σ_v for all v' that occurs in the averaging.

1. and 2. in this definition contain the requirement that the hidden theory be capable of reproducing the results yielded by T, and 3. demands of T' that it be "more deterministic", or "less statistical" than the theory whose hidden theory it is. This latter demand is not explicit in the hidden variable definitions occuring in the literature; however, in most concrete cases it is assumed that the hypothetical hidden variable theory of quantum mechanics can eliminate *all* uncertainty inherent in quantum mechanical state description, where the uncertainty is usually measured by dispersions of quantum states.

It will be seen below that dispersion-freeness is an extremely strong requirement, which weakens the significance of the "impossibility proofs" considerably. It should be emphasized that the constraint put on the hidden theory T' in the Definition 9.1 is much weaker: it is not required of the hidden theory (though it is allowed) that it be capable of eliminating completely all uncertainty inherent in T: only reducing the uncertainty by T' is required; furthermore, this latter requirement is formulated relative to two "variables": the averaging and the measure of uncertainty, precise form and properties of both of which have deliberately been left unspecified in the above definition. One obtains a more specific definition that can be the basis of provable mathematical statements just by making both the concept of averaging precise and choosing particular measures of uncertainty σ_v and $\sigma_{v'}$. This can only be done in a fixed, well defined framework of quantum mechanics. In what follows, the definition of hidden theory will be given within the operator algebraic framework of quantum mechanics.

Let $\mathcal{A}, \mathcal{B}$ be unital C^*-algebras, and denote by $E(\mathcal{A}), E(\mathcal{B})$ their state spaces. A pair $(\mathcal{A}, E(\mathcal{A}))$ will be called *physical theory* for short. Recall that the dispersion of a state φ on an element $A \in \mathcal{A}$ is, by definition

$$\sigma_\varphi(A) = \varphi(A^2) - |\varphi(A)|^2 \geq 0$$

(this is just the operator algebraic reformulation of the Definitions 2.23 and 2.25).

For a given $\psi \in E(\mathcal{B})$ let $M_\psi(E(\mathcal{B}))$ denote the set of all normalized Borel measures μ on $E(\mathcal{B})$ that have ψ as their barycenter, i.e. which decompose ψ in the sense of the following integral

$$\psi(B) = \int \omega(B) d\mu(\omega) \qquad B \in \mathcal{B} \tag{9.1}$$

Proposition 9.1 *$M_\psi(E(\mathcal{B}))$ is nonempty for any ψ because it contains one element at least, namely it always contains the Dirac measure δ_ψ concentrated at ψ; furthermore, δ_ψ is the only element in $M_\psi(E(\mathcal{B}))$ if ψ is a pure state.*

Proof: If there is another element μ in $M_\psi(E(\mathcal{B}))$ different from δ_ψ, then $\text{supp}(\mu)$ contains two different points that can be separated by two Borel sets E_1 and E_2 such that

$$E_1 \cup E_2 = E(\mathcal{B}) \qquad \mu(E_1) \neq 1, 0 \qquad \mu(E_2) \neq 1, 0$$

and then ψ_1 and ψ_2 defined by

$$\psi_1 = (\mu(E_1))^{-1} \int_{E_1} \omega d\mu(\omega)$$
$$\psi_2 = (\mu(E_2))^{-1} \int_{E_2} \omega d\mu(\omega)$$

are different states, furthermore

$$\psi = \mu(E_1)\psi_1 + \mu(E_2)\psi_2$$

thus ψ can be written as a non-trivial combination of two distinct states, i.e. ψ is not a pure state. □

The linear map $L: \mathcal{B} \to \mathcal{A}$ is called *positive* if $0 \leq B$ implies $0 \leq LB$, and L is said to be *unit preserving* if it carries the unit element I of $\mathcal{B}$ into the unit of $\mathcal{A}$. In what follows, all positive maps will be supposed to be unit preserving, unless stated otherwise explicitly. $\mathcal{D}(\mathcal{B}, \mathcal{A})$ denotes the set of all positive, unit preserving maps from $\mathcal{B}$ into $\mathcal{A}$. If $L \in \mathcal{D}(\mathcal{B}, \mathcal{A})$, then

$$(L^*\varphi)(B) = \varphi(LB)$$

defines a state $L^* \in E(\mathcal{B})$.

For a positive, unit preserving map $L: \mathcal{B} \to \mathcal{A}$ the following "generalized Cauchy-Schwartz inequality" holds holds for all selfadjoint $x \in \mathcal{B}$.

$$L(x^2) \geq L(x)^2 \qquad X \in \mathcal{B} \tag{9.2}$$

We can now give the key definition of this chapter.

Definition 9.2 Let $(\mathcal{A}, E(\mathcal{A}))$ and $(\mathcal{B}, E(\mathcal{B}))$ be two physical theories. $(\mathcal{B}, E(\mathcal{B}))$ is a said to be a *hidden theory* of $(\mathcal{A}, E(\mathcal{A}))$ with respect to an L if the following hold

(1) L is a linear, positive, unit preserving map from $\mathcal{B}$ into $\mathcal{A}$
(2) for all $\varphi \in E(\mathcal{A})$ there is a $\mu \in M_{L^*\varphi}(E(\mathcal{B}))$ measure such that
(2.1) for every $B \in \mathcal{B}$ for which LB is selfadjoint and $\sigma_\varphi(LB) > 0$ we have that

$$\sigma_\varphi(LB) > \sigma_\omega(B) \qquad \text{for all} \quad \omega \in \text{supp}(\mu)$$

(2.2) for every $B \in \mathcal{B}$ for which LB is selfadjoint and $\sigma_\varphi(LB) = 0$ we have that

$$\sigma_\varphi(LB) = \sigma_\omega(B) = 0 \qquad \text{for all} \quad \omega \in \text{supp}(\mu)$$

Definition 9.2 is a specification of Definition 9.1 within the operator algebraic framework of quantum mechanics. Formally, von Neumann's original definition [168] can be obtained as a special case of Definition 9.2 by chosing $\mathcal{A} = \mathcal{B} = \mathcal{B}(\mathcal{H})$ and taking L=identity, where $\mathcal{B}(\mathcal{H})$ is the C^*-algebra of all bounded operators on the Hilbert space $\mathcal{H}$.

Thirty two years after von Neumann, the hidden variable problem was re-investigated by Misra [101] in algebraic quantum mechanics. Misra chose $\mathcal{A} = \mathcal{B}$ and L=identity, but he did not assume $\mathcal{A}$ to be $\mathcal{B}(\mathcal{H})$ on some $\mathcal{H}$, rather, in [101] $\mathcal{A}$ was an arbitrary C^*-algebra. However, both Misra and von Neumann demanded that the states be decomposable into *dispersion-free* states in the manner (2.1).

Von Neumann's result is well known: there is no dispersion-free state on $\mathcal{B}(\mathcal{H})$ if $\mathcal{H}$ is the infinite dimensional, separable, complex Hilbert space occuring in the description of quantum systems. As we mentioned in the historical introduction to this chapter, von Neumann's impossibility proof was criticized very soon after it had appeared. Misra was aware of those critiques, and, in his opinion, the essence of the critique raised was the charge that, by assuming the *linearity* of states, especially the linearity of states on incompatible observables, von Neumann tacitly assumed the universal validity of the Heisenberg uncertainty relation, thus it was neither surprising, nor significant (from the point of view of the hidden variable problem) that von Neumann's assumption excluded the existence of dispersion-free states. In short, according to Misra, it was the linearity assumption that made the charge of "circular reasoning" possible. Misra also knew Gleason's result, which showed that von Neumann's assumption on linearity is not as strong as it may seem; however, Misra investigated the hidden variable problem on a general C^*-algebra, and since it was not known at that time whether Gleason's theorem is valid for von Neumann algebras that are more general than $\mathcal{B}(\mathcal{H})$, he could not refer to a general Gleason's theorem if he wanted to make his impossiblity proof immune against the same charge that could be raised against von Neumann's proof.

Misra saw the solution of this problem in weakening of the definition of state: instead of assuming the "states" to be positive, linear functionals, he allowed states to be not necessarly linear, *monotone positive* functionals on the C^*-algebra in the sense of the following definition.

Definition 9.3 A map $\varphi: \mathcal{A} \to \mathbb{C}$ from a C^*-algebra $\mathcal{A}$ into the complex numbers is *monotone positive* if it takes on real values on selfadjoint elements and if $0 \leq X \leq Y$ implies $0 \leq \varphi(X) \leq \varphi(Y)$.

Misra succeeded in proving that every state on $\mathcal{A}$ can be obtained as an average over dispersion-free, monotone positive functionals on $\mathcal{A}$ if and only if $\mathcal{A}$ is a commutative C^*-algebra. (Theorem 2 in [101]).

As noted already, the dispersion freeness is an extremely strong demand both physically as well as mathematically, even if one tries to weaken the definition of state. One can argue, for instance, that it can never be decided empirically whether a physical state is strictly dispersion-free or not because of the presence of non-zero measurement error unavoidable in every real physical measurement; consequently, any hidden variable (or hidden theory) definition formulated in terms of dispersion-free states is physically too restrictive. Thus, emphasizing the extreme restrictivity of dispersion-freeness and the unreasonabilty of demanding it as a condition to be met by a hidden theory means a much more serious critique of von Neumann's (and oher's) impossiblity proof than the usual charge of circularity of reasoning. The trouble with dispersion-freeness is not that it makes the impossiblity proof circular because, in a sense, circularity is a sign of correctness: a proof, a valid mathematical proof that is, together with the statement it proves, is but making transparent something that is implicit only in the hypothesis. The dispersion-freeness is simply too restrictive a demand, and a natural way to strengthen the impossibility proofs is to weaken the definition of hidden theory by giving up dispersion-freeness and by replacing it with a more reasonable demand. Since one expects the hidden theory to decrease uncertainty, the requirement that the hidden states have pointwise strictly less dispersion, i.e conditions 2.1 and 2.2 in Definition 9.2 seem to be the weakest ones we must insist on.

Also, the assumption L=identity, which was made (implicitly) by both von Neumann and Misra, is by no means necessary mathematically, and it can hardly be justified on physical grounds. The advantage of introducing L as a variable into the Definition 9.2 is that L serves as a mathematical tool to investigate what structural properties of the observable quantities are responsible for a given uncertainty content embodied in the description by that structure of the observable world. To put this into more mathematical terms, consider the category of C^*-algebras with the positive maps as morphisms. If, in addition to positivity, L preserves a certain algebraic structure, and one can prove a *negative* result on the problem whether a hidden theory $(\mathcal{B}, E(\mathcal{B}))$ via L of a given theory $(\mathcal{A}, E(\mathcal{A}))$ exists (in the sense of Definition 9.2), then one singles out a subcateory consisting of C^*-algebras whose algebraic structure represent a given uncertainty in description of the physical system.

Thus, in the interpretation based on Definition 9.2, the research in the field of hidden variables is not a philosophically motivated physical search for a new physical theory that could replace quantum mechanics and which can provide a deterministic description of the physical world; nor is it a philosophical research that one should do in the hope that by finding those hypothetical variables or by proving that they are not impossible one can show that we are not forced to consider certain metaphysical principles as incompatible with physics; nor should negative results stating the nonexistence of hidden theories be considered as a proof of invalidity of determinism or realism, and such theorems should by no means be viewed as reason enough to abandon those metaphysical (and very convenient) principles. The aim of the hidden variable investigations is more modest, less ambitious and can be sharply formulated: On the present interpretation the aim of the hidden variable investigations is to characterize the role of algebraic structures from the point of view of the uncertainty content of statistical theories applying noncommutative probability theory.

9.2.2. NEGATIVE RESULTS ON DISPERSIVE HIDDEN THEORIES

Proposition 9.2 *Assume that $\mathcal{A}$ contains at least one element different from the unit I of $\mathcal{A}$. Then there is a dispersion-free state φ in $E(\mathcal{A})$ if and only if there is a non-trivial, two-sided ideal J in $\mathcal{A}$ such that the quotient algebra $\mathcal{A}/J$ is commutative.*

Proof: Let φ be a dispersion-free state. We claim that

$$J = \{X \in \mathcal{A} \mid \varphi(X) = 0\}$$

is a non-trivial two-sided ideal such that $\mathcal{A}/J$ is commutative. Obviously, J is a linear and selfadjoint set, furthermore, it also is non-trivial because if X is an arbitrary element different from I, then $X - \varphi(Y)I \neq I$, but $\varphi(Y - \varphi(Y)I) = 0$, i.e. $(Y - \varphi(Y)I)$ is a non-trivial element in J. J also is an ideal because every element in $\mathcal{A}$ can be obtained as linear combination of selfadjoint elements, and if $X \in \mathcal{A}$ is selfadjoint, then for all $Y \in J$ we have by the Cauchy-Schwarz inequality and by the dispersion freeness of φ

$$|\varphi(XY)| \leq \varphi(XX^*)\varphi(YY^*) = \varphi(X)^2\varphi(Y)^2 = 0$$

Thus $(XY) \in J$ and, for similar reasons, $(YX) \in J$. Finally, if $\{Y\}$ denotes the equivalence class of Y in $\mathcal{A}/J$, then the map h defined by $\{y\} \to h(\{y\}) = \varphi(Y)$ is an isomorphism between $\mathcal{A}/J$ and the complex numbers, because if $\varphi(Y) = \varphi(Z)$, then $\varphi(Y - Z) = 0$, thus $\{Y\} = \{z\}$, consequently h is invertible; furthermore, by

$$h(\{Y\}\{Z\}) = h(\{YZ\}) = \varphi(YZ) = \varphi(Y)\varphi(Z)$$

h also preserves the product. Conversely, if there is a J ideal such that $\mathcal{A}/J$ is commutative, then there is a dispersion-free state over $\mathcal{A}/J$ (all pure states are dispersion-free). □

As a corollary of the above proposition we have

Proposition 9.3 *There is no dispersion-free state on $\mathcal{A}$ if $\mathcal{A}$ is a simple C^*-algebra.*

Proposition 9.4 *$(\mathcal{A}, E(\mathcal{A}))$ is a dispersion-free hidden theory of itself (i.e. with respect to the identity map as L) if and only if $\mathcal{A}$ is a commutative C^*-algebra.*

Proof: If $\mathcal{A}$ is commutative, then by the Gelfand representation theorem it is isomorphic to $C(\Gamma)$, the space of continuous functions over the characters Γ of $\mathcal{A}$. Thus every $\varphi \in E(\mathcal{A})$ can be considered as a linear functional over $C(\Gamma)$. By the Riesz representation theorem, for such a functional a μ measure on Γ can be given with the help of which φ can be obtained in the form $\varphi(X) = \int f_x d\mu$ (f_x is the function in $C(\Gamma)$ that represents the element

$X \in \mathcal{A}$). Clearly, all the states in suppμ are dispersion-free because every character is dispersion-free. Conversely, let us assume that $(\mathcal{A}, E(\mathcal{A}))$ is a dispersion-free hidden theory of itself. Let X and Y be elements satisfying $X > Y \geq 0$. For an arbitrary state φ there is then a μ measure on $E(\mathcal{A})$ such that

$$\varphi(X^2) = \int \omega(X^2) \mathrm{d}\mu(\omega) = \int \omega(X)^2 \mathrm{d}\mu(\omega)$$
$$\varphi(Y^2) = \int \omega(Y^2) \mathrm{d}\mu(\omega) \int \omega(Y)^2 \mathrm{d}\mu(\omega)$$

It follows that

$$\varphi(X^2) - \varphi(Y^2) = \int (\omega(X)^2 - \omega(Y)^2) \mathrm{d}\mu(\omega) \geq 0 \qquad (9.3)$$

because $\omega(X) \geq \omega(Y)$. Since (9.3) holds for an arbitrary state φ, it follows that $X^2 \geq Y^2$. The statement is now the consequence of the fact that if for every pair of elements X, Y in a C^*-algebra $\mathcal{A}$ such that $X \geq Y$ the relation $X^2 \geq Y^2$ also holds, then $\mathcal{A}$ is commutative. $\square$

Proposition 9.5 *Let $\mathcal{A}$ be a simple C^*-algebra. Then $(\mathcal{B}, E(\mathcal{B}))$ is not a hidden theory of $(\mathcal{A}, E(\mathcal{A}))$ with respect to an $L \in \mathcal{D}(\mathcal{B}, \mathcal{A})$ if there is a $\varphi' \in E(\mathcal{A})$ state such that $L^*\varphi' \in E(\mathcal{B})$ is a pure state.*

Proof: Let us assume that $(\mathcal{B}, E(\mathcal{B}))$ is a hidden theory of $(\mathcal{A}, E(\mathcal{A}))$ with respect to L. Then there is a $\mu \in M_{L^*\varphi}(E(\mathcal{B}))$ measure for an arbitrary state $\varphi \in E(\mathcal{A})$ such that

$$\varphi(LB) = \int \omega(B) \mathrm{d}\mu(\omega) \qquad B \in \mathcal{B} \qquad (9.4)$$

$$\sigma_\varphi(LB) \geq \sigma_\omega(B) \qquad \omega \in \mathrm{supp}\mu, \quad LB \text{ selfadjoint} \qquad (9.5)$$

For fixed B both sides of (9.5) define a function on $E(\mathcal{B})$ that is integrable with respect to μ. Integrating both sides of (9.5) one obtains

$$\varphi((LB)^2 - LB^2) \geq (\varphi(LB))^2 - \int \omega(B)^2 \mathrm{d}\mu(\omega) \qquad (9.6)$$

The left hand side of (9.6) is not positive for all $B \in (\mathcal{B})_{sa}$ and for all $\varphi \in E(\mathcal{A})$ by the generalized Cauchy-Schwarz inequality for positive maps. In particular, (9.6) also holds for φ'. Since $L^*\varphi'$ is a pure state, μ is the Dirac measure $\delta_{L^*\varphi'}$ by Proposition 9.1. Since $\mathcal{A}$ is simple, φ' cannot be dispersion-free by Proposition 9.2, thus there is a $B_0 \in \mathcal{B}_{sa}$ such that $\sigma_{\varphi'}(LB_0) > 0$;

therefore, substituting φ' and B_0 into (9.6), this (9.6) inequality must hold as a strict inequality by the definition of hidden theory, i.e. we must have

$$0 \geq \varphi'((LB_0)^2) - LB_0^2) > (\varphi'(LB_0))^2 - \int (\omega(B_0))^2 d\delta_{L^*\varphi'}(\omega) = 0 \quad (9.7)$$

which is not possible. □

Definition 9.4 A map $L \in \mathcal{D}(\mathcal{B}, \mathcal{A})$ is called *Jordan homomorphism* if it preserves the anticommutator

$$[X, Y]_+ = XY + YX$$

in the sense

$$L[X, Y]_+ = [LX, LY]_+ \quad (9.8)$$

Let $[\varphi]$ denote the set of states having the form $\sum_j^n r_j \varphi_j$, where

$$\varphi_j(X) = \frac{\varphi(Y_j^* X Y_j)}{\varphi(Y_j^* Y_j)}$$

whenever $\varphi(Y_j^* Y_j) \neq$ and for $Y_j \in \mathcal{A}$ and $r_j > 0$ such that $\sum_j^n r_j = 1$. It is known [143] that $L \in \mathcal{D}(\mathcal{B}, \mathcal{A})$ is a Jordan homomorphism if and only if for all pure states $\varphi \in E(\mathcal{A})$ the states $L^*\varphi \in E(\mathcal{B})$ are pure, too, and $L^*[\varphi] = [L^*\varphi]$. Since the set of pure states on a C^*-algebra is non-empty, the following Proposition is consequence of Proposition 9.5.

Proposition 9.6 *Let $\mathcal{A}$ be a simple C^*-algebra. Then there is no physical theory $(\mathcal{B}, E(\mathcal{B}))$ which is a hidden theory of $(\mathcal{A}, E(\mathcal{A}))$ with respect to an L Jordan homomorphism.*

Proposition 9.6 tells us that if $(\mathcal{B}, E(\mathcal{B}))$ is a hidden theory of $(\mathcal{A}, E(\mathcal{A}))$ with respect to an L, then L must distinguish the Jordan algebra structures of $\mathcal{A}$ and $\mathcal{B}$; that is, in this case the Jordan algebra structures of $\mathcal{A}$ and $\mathcal{B}$ must be regarded as different. One can also formulate this by saying that the C^*-algebras having similar Jordan algebra structure belong to the subcategory representing "at least the same uncertainty".

This is something to be expected because the basic idea of algebraic quantum mechanics is that it is the Jordan algebra structure of observables that bears physical meaning only, thus one expects intuitively that similar Jordan algebra structures cannot be radically different, and indeed, as Proposition 2.4 shows, they are not very different if considered from the point of view of the uncertainty displayed by the dispersion of states defined over them. Thus Proposition 9.6 also expresses a kind of consistency of the definition of hidden theory with the basic idea of algebraic quantum mechanics.

However, the onto Jordan homomorhisms are not the only types of positive maps whose dual map carries a pure state into a pure state. The following definition is due to Störmer, [142].

Definition 9.5 Assume that $\mathcal{A}$ acts on a Hilbert space $\mathcal{H}$, $\mathcal{A} = \mathcal{A}(\mathcal{H})$. A map $L \in \mathcal{D}(\mathcal{B}, \mathcal{A}(\mathcal{H}))$ is called "Class 0" if $L^*\varphi$ is pure for all pure $\varphi \in E(\mathcal{A}(\mathcal{H}))$, and L is called "Class 1" if $L^*\varphi_0$ is pure for all *vector* states φ_η.

The onto Jordan homomorphisms are "Class 0" obviously, and a complete characterisation of "Class 1" maps is known:

Proposition 9.7 (Störmer's Theorem) $L \in \mathcal{D}(\mathcal{B}, \mathcal{A}(\mathcal{H}))$ *is "Class 1" if and only if L is of the form*

$$L = V^*h(\bullet)V$$

where $V: \mathcal{H} \to \mathcal{K}$ is a linear isomorphism from $\mathcal{H}$ into the Hilbert space $\mathcal{K}$, and h is an irreducible $*$-homomorphism (or anti-homomorphism) from $\mathcal{B}$ into $\mathcal{B}(\mathcal{K})$.

Thus Proposition 9.5 implies the following

Proposition 9.8 *If $\mathcal{A}(\mathcal{H})$ is simple, then $(\mathcal{B}, E(\mathcal{B}))$ is not a hidden theory of $(\mathcal{A}(\mathcal{H}), E(\mathcal{A}(\mathcal{H})))$ with respect to a "Class 1" L.*

This proposition can be interpreted as showing that the perturbations of homomorphisms (or anti-homomorphisms) described in Störmer's theorem (Proposition 9.7) do not take us out from the subcategory formed by C^*-algebras that represent a given uncertainty in description of physical systems, that is to say, the subcategory is stable with respect to the perturbations specified in Proposition 9.7.

Another important class of positive maps are formed by *conditional expectations* (also called: projections of norm one): Let $\mathcal{A}$ be a C^*-subalgebra of $\mathcal{B}$ containing the unit element of $\mathcal{B}$. Recall that the map $L \in \mathcal{D}(\mathcal{B}, \mathcal{A})$ is called then a conditional expectation from $\mathcal{B}$ onto $\mathcal{A}$ if the restriction of L onto $\mathcal{A}$ is the identity map (see the Chapter 8, especially Section 8.2). If, however, L is a conditional expectation, then it is natural to weaken slightly the definition of hidden theory by replacing 2.1 and 2.2 in Definition 9.1 by the following 2.1 (CE) and 2.2 (CE) requirements:

2.1 (CE) for all $X \in \mathcal{B}$ which is not contained in $\mathcal{A}$ and for which $LX \in \mathcal{A}_{sa}$ we have that

if
$$\sigma_\varphi(LX) > 0$$
then
$$\sigma_\varphi(LX) > \sigma_\omega(X)$$

if
$$\sigma_\varphi(LX) = 0$$
then
$$\sigma_\varphi(LX) = \sigma_\omega(X)$$

2.2 (CE) for all $X \in \mathcal{B}$ which also is contained in $\mathcal{A}$ and for which $LX \in \mathcal{A}_{sa}$ we have that
$$\sigma_\varphi(LX) \geq \sigma_\omega(X)$$
and these conditions hold for all $\omega \in \text{supp}\mu$.

The meaning of this modification of the Definition 9.2 is that in this "conditional expectation case" we only require of the hidden theory that it decrease the dispersion on the elements in $\mathcal{B}$ which do *not* belong to $\mathcal{A}$, and on elements in $\mathcal{B}$ that also lie in $\mathcal{A}$ we only demand that it does not increase the dispersion. Clearly, without this modification Definition 9.2 would require of $(\mathcal{A}, E(\mathcal{A}))$ in the case of L=conditional expectation that it be a hidden theory of itself. In the next proposition "hidden theory" is meant in the slightly weakened sense just described. Before stating the proposition recall that an $\mathcal{A}$ algebra is *called uniformly hyperfinite* (UHF algebra) if there is a series of full matrix algebras M_j ($j = 1, 2, \ldots$) such that all M_j have a common unit and the union $\cup_j M_j$ is dense in norm in $\mathcal{A}$.

Proposition 9.9 *Let $\mathcal{A}$ be a C^*-subalgebra of $\mathcal{B}$ containing the unit of $\mathcal{B}$ and the full matrix algebra of complex two by two matrices M_2, (unit of which need not be the unit of $\mathcal{A}$). Then $(\mathcal{B}, E(\mathcal{B}))$ is not a hidden theory of $(\mathcal{A}, E(\mathcal{A}))$ with respect to an $L \in \mathcal{D}(\mathcal{B}, \mathcal{A})$ conditional expectation. In particular, $(\mathcal{B}, E(\mathcal{B}))$ is not a hidden theory of $(\mathcal{A}, E(\mathcal{A}))$ with respect to a conditional expectation $L \in \mathcal{D}(\mathcal{B}, \mathcal{A})$ if $\mathcal{A}$ is a UHF algebra.*

Proof: In the proof we shall use the following property of a conditional expectation L:
$$LXBY = LXLBLY \quad \text{for all} \quad X, Y \in \mathcal{A}; B \in \mathcal{B}$$

In particular one has
$$L[X, B] = [X, LY] \quad X \in \mathcal{A}, Y \in \mathcal{B} \tag{9.9}$$

Consider now the matrices $S_i = (1/2)P_i$ ($i = 1, 2, 3$), where P_i are the two dimensional Pauli matrices, and assume that $(\mathcal{B}, E(\mathcal{B}))$ is a hidden theory of $(\mathcal{A}, E(\mathcal{A}))$ with respect to an L conditional expectation. Let $\varphi \in E(\mathcal{A})$ be the state whose restriction to M_2 is given by the density matrix

$$\begin{pmatrix} 1 & 1 \\ 1 & 1 \end{pmatrix}$$

An explicit elementary calculation shows the validity of the following inequality

$$\sigma_\varphi(S_2)\sigma_\varphi(S_3) = (1/4)\,|\,\varphi([S_2, S_3])\,|^2 \qquad (9.10)$$

Since the uninteresting case $\mathcal{A} = \mathcal{B}$ can be excluded, there is an $s_2 \in \mathcal{B}_{sa}$ not belonging to $\mathcal{A}$ such that $Ls_2 = S_2$, and then by the definition of hidden theory we must have

$$\sigma_\varphi(S_2) > \sigma_\omega(s_2) \qquad (9.11)$$
$$\sigma_\varphi(S_3) \geq \sigma_\omega(S_3) \qquad (9.12)$$
$$\text{for all} \qquad \omega \in \mathrm{supp}\mu \qquad (9.13)$$

where $\varphi(LX) = \int \omega(X)\mathrm{d}\mu(\omega)$ is a hidden decomposition of the state φ. It follows that

$$\sigma_\varphi(S_2)\sigma_\varphi(S_3) > \sigma_\omega(s_2)\sigma_\omega(s_3) \quad (\omega \in \mathrm{supp}\mu)$$

By the Heisenberg uncertainty relation – which is valid in every C^*-algebra – we have

$$\sigma_\psi(s_2)\sigma_\psi(s_3) \geq (1/4)\,|\,\psi[s_2, s_3]\,|^2$$

holds for every $\psi \in E(\mathcal{B})$, consequently also for all $\omega \in \mathrm{supp}\mu$, thus

$$\sigma_\varphi(S_2)^{1/2}\sigma_\varphi(S_3)^{1/2} > (1/2)\,|\,\omega([s_2, S_3])\,| \qquad (\omega \in \mathrm{supp}\mu) \qquad (9.14)$$

Integrating both sides of (9.14) with respect to μ, and using (9.9) one obtains

$$\sigma_\varphi(S_2)^{1/2}\sigma_\varphi(S_3)^{1/2} > (1/2)\int |\,\omega([s_2, S_3])\,|\,\mathrm{d}\mu(\omega)$$
$$\geq |\int \omega([s_2, S_3])\mathrm{d}\mu\,| = (1/2)\,|\,\varphi(L[s_2, S_3])\,|$$
$$= (1/2)\,|\,\varphi([Ls_2, LS_3])\,| = (1/2)\,|\,\varphi([S_2, S_3])\,|$$

And this last inequality contradicts (9.10). $\square$

If $\mathcal{A}$ is non-commutative, then it "contains M_2", and "$\mathcal{A}$ contains M_2" in the above proposition can be replaced by "$\mathcal{A}$ is non-commutative".

Thus Proposition (9.9) tells us that a non-commutative physical theory cannot be embedded into a larger one so that the latter one be a hidden theory of the former one. One can also say that a non-commutative physical theory cannot be transformed into a hidden theory by enlarging, by adding new elements to it. Another wording of interpretation of Proposition 9.9 can be that non-commutative C^*-algebras embedded into each other via conditional expectations form a subcategory of the same uncertainty.

We close this section by commenting on the problem of "approximate hidden variables".

The problem of "approximate hidden variables" or "approximate dispersion-free states" was first raised by Jauch and Piron in [74]. They formulated the problem in the lattice theoretical framework of quantum mechanics as follows. Let $\mathcal{L}$ be an orthomodular σ-lattice and $E(\mathcal{L})$ be the set of states over $\mathcal{L}$. The pair $(\mathcal{L}, E(\mathcal{L}))$ is said to admit approximate dispersion-free states if there is a sequence of states $\varphi_n \in E(\mathcal{L})$ such that the overall dispersions defined by

$$\sigma_{\varphi_n} \equiv = \sup(\varphi_n(X) - \varphi_n(X)^2)$$

(the supremum taken over all $X \in \mathcal{L}$) tend to zero for $n \to \infty$.

Whereas the Hilbert lattice of projections of the Hilbert space of ordinary quantum mechanics was found by Jauch and Piron not to admit approximate hidden variables, they emphasized that nothing was known for more general quantum propositional systems.

Later on, the approximate hidden variable problem was reformulated by Misra [101] within the operator algebraic approach to quantum mechanics. Misra defines the pair $(\mathcal{A}, E(\mathcal{A}))$ to admit approximate hidden variables if for all $\epsilon > 0$ every state $\varphi \in E(\mathcal{A})$ can be obtained in the form

$$\varphi(A) = \int \omega_t^\epsilon(X) d\mu(t) \qquad X \in \mathcal{A}$$

where μ is a positive, normalized measure on some set Ω, and ω_t^ϵ are functionals defined on $\mathcal{A}$ for all t in Ω such that the overall dispersions defined by

$$\sigma_{\omega_t^\epsilon} \equiv = \sup(\omega_t^\epsilon(A^2) - (\omega_t^\epsilon(A)^2))$$

(the supremum taken over the elements $A \in \mathcal{A}$ with norm less than, or equal to one) satisfy $\sigma_{\omega_t^\epsilon} \leq \epsilon$ for all t. As Misra noted, none of the known impossibility proofs regarding the hidden variables was strong enough to exclude the existence of approximate hidden variables in the above sense.

The appearance of these approximate hidden variable definitions is a clear sign of recognition that the assumption of dispersion-freeness in the hidden variable definitions is much too restrictive for both practical and theoretical purposes. One wanted to have a hidden variable definition where this too stringent condition is relaxed.

A natural adaptation of the idea of approximate hidden variables in terms of the algebraic hidden theory concept analyzed in this chapter is the following definition.

Definition 9.6 The physical theory $(\mathcal{B}, E(\mathcal{B}))$ is an *approximate hidden theory* of $(\mathcal{A}, E(\mathcal{A}))$ with respect to $L \in D(\mathcal{B}, \mathcal{A})$ if, given a state $\phi \in E(\mathcal{A})$, for every $\epsilon > 0$ there exists a $\mu^\epsilon \in M_{L^*\phi}(E(\mathcal{B}))$ such that

$$\sigma_{\omega^\epsilon} < \epsilon$$

for all $\omega^\epsilon \in \mathrm{supp}\mu^\epsilon$, where σ_{ω^ϵ} is the overall dispersion of the state ω defined by

$$\sigma_{\omega^\epsilon} \equiv= \sup_\omega(\omega(A^2) - (\omega(A)^2))$$

The Propositions 9.5, 9.6, 9.8 and 9.9 clearly imply then that under their assumptions there does not exist an approximate hidden theory of a physical theory $(\mathcal{A}, E(\mathcal{A}))$.

9.3. No-go results on entropic hidden theories

In translating the Definition 9.1 into Definition 9.2 we have used the dispersion as a measure of uncertainty of quantum states. However, the dispersion is not always an appropriate measure of uncertainty; in particular, there are strong arguments against the adequacy of dispersion as a measure of uncertainty understood as measurement error or even of an uncertainty reflecting the measurement situation in any reasonable sense. On the other hand, one can define quantities that can be viewed as uncertainties associated with a measurement situation: these are the entropic uncertainties that depend not only on a state and on the observable but also on the partition of real numbers into disjoint sets, which can then be interpreted as the resolution of the measuring device. The definition of entropic uncertainty was given in Definition 2.7 in the Hilbert space quantum mechanics, but the definition makes perfect sense also in terms of von Neumann algebras:

Definition 9.7 Let $\{d_i\}$ be a partition of real numbers into disjoint Borel measurable sets, and let X is a selfadjoint element in the von Neumann algebra $\mathcal{M}$. The *entropic uncertainty* of the state ϕ on X with respect to the partition is given by

$$S(\phi, X, \{d_i\}) \equiv -\sum_i^\infty \phi(\mathrm{P}^X(d_i)) \log \phi(\mathrm{P}^X(d_i)) \qquad (9.15)$$

THE PROBLEM OF HIDDEN VARIABLES 157

If we interpret the partition $\{d_i\}$ as the resolution of the measuring device that measures X, the quantity $S(\phi, X, \{d_i\})$ is sensitive to the measurement setup. It is a natural idea to try to replace dispersion in Definition 9.2 of hidden theory by the entropic uncertainty and to investigate the problem of existence of entropic hidden theories.

Definition 9.8 Let $\mathcal{M}, \mathcal{N}$ be von Neumann algebras and $(\mathcal{N}, E(\mathcal{N}))$ and $(\mathcal{M}, E(\mathcal{M}))$ be the corresponding physical theories. $(\mathcal{N}, E(\mathcal{N}))$ is said to be an *entropic hidden theory* of $(\mathcal{M}, E(\mathcal{M}))$ with respect to $L \in D(\mathcal{N}, \mathcal{M})$, if the measure μ in Definition 9.2 can be chosen such that instead of 2.1 and 2.2 in Definition 9.2 the following 2.1' and 2.2' conditions are fulfilled: For any finite partition $\{d_i\}$ and for all selfadjoint LX and X one has

$$S(\phi, LX, \{d_i\}) > S(\omega, X, \{d_i\}) \quad \text{if} \quad S(\phi, LX, \{d_i\}) > 0 \quad (9.16)$$
$$S(\phi, LX, \{d_i\}) = S(\omega, X, \{d_i\}) \quad \text{if} \quad S(\phi, LX, \{d_i\}) = 0 \quad (9.17)$$

for all $\omega \in \operatorname{supp}\mu$

Similarly to the case of dispersive hidden theories, a statement asserting the nonexistence of entropic hidden theory of a physical theory with respect to a positive map L preserving a certain algebraic structure can be interpreted as the determination of a certain algebraic structure without restriction of which the measurement uncertainties cannot be reduced, i.e. such negative statements single out the "at least the same measurement uncertainty" subcategory in the category of C^*-algebras.

Proposition 9.10 *The physical theory $(\mathcal{N}, E(\mathcal{N}))$ is not an entropic hidden theory of $(\mathcal{M}, E(\mathcal{M}))$ with respect to $L \in D(\mathcal{N}, \mathcal{M})$ if L is a Jordan homomorphism.*

Proof: Let us assume that $(\mathcal{N}, E(\mathcal{N}))$ is an entropic hidden theory of $(\mathcal{M}, E(\mathcal{M}))$ with respect to an L Jordan homomorphism and let

$$\phi(LX) = \int \omega \, d\mu(\omega) \quad (9.18)$$

be a hidden decomposition of the arbitrarily chosen state $\phi \in E(\mathcal{M})$. Then, if LX and X are selfadjoint elements, we have

$$-\sum_i \phi(\mathrm{P}^{LX}(d_i)) \log \phi(\mathrm{P}^{LX}(d_i)) \geq -\sum_i \omega(\mathrm{P}^X(d_i)) \log \omega(\mathrm{P}^X(d_i)) \quad (9.19)$$

for all $\omega \in \operatorname{supp}\mu$ and for every finite partition $\{d_i\}$. Integrating (9.18) with respect to μ one obtains

$$-\sum_i \phi(\mathrm{P}^{LX}(d_i)) \log \phi(\mathrm{P}^{LX}(d_i)) \geq -\sum_i \int \omega(\mathrm{P}^X(d_i)) \log \omega(\mathrm{P}^X(d_i)) \, d\mu(\omega)$$
$$(9.20)$$

Let us pick a pure state ϕ. Then $L^*\phi$ also is a pure state, since L is a Jordan homomorphism; thus, by Proposition 9.1 $\mu = \delta_{L^*\phi}$. Substituting this μ into (9.20) one obtains

$$-\sum_i \phi(\mathrm{P}^{LX}(d_i)) \; \log \; \phi(\mathrm{P}^{LX}(d_i)) \geq \qquad (9.21)$$

$$-\sum_i \phi(L\mathrm{P}^X(d_i)) \; \log \; \phi(L\mathrm{P}^X(d_i)) \mathrm{d}\mu(\omega) \qquad (9.22)$$

Let now LX be an arbitrary selfadjoint element. There is a pure state ϕ such that
$$\phi((LX)^*LX) = \|LX\|^2$$
which implies that there is non-trivial projector E in N such that $0 < \phi(LE) < 1$. Since L is a Jordan homomorphism, it follows that LE also is a non-trivial projector. The spectral resolutions of E and LE are given by

$$E = E + 0(I - E)$$
$$LE = LE + 0(LI - LE)$$

thus if we evaluate (9.21) for the two-valued observable LE taking the partition
$$d_1 = \{1\} \quad d_2 = \{0\} \quad d_3 = \mathbb{R} \setminus \{0, 1\}$$
we obtain

$$-\phi(LE)\log(\phi(LE) - \phi(LI - LE) \; \log \; \phi(LI - LE) \geq \qquad (9.23)$$
$$-\phi(LE)\log(\phi(LE) - \phi(L(I - E)) \; \log \; \phi(L(I - E)) \qquad (9.24)$$

that is, the inequality (9.19) holds as an equality; however, by the definition of entropic hidden theory, it should hold as a strict inequality because $0 < \phi(LE) < 1$ implies that the left hand side of (9.19) is greater than zero. □

Formally, this proposition is very similar to Proposition 9.6, but what it tells us is conceptually different from what Proposition 9.6 does. First of all, this proposition shows the a priori not obvious consistency of the definition of hidden theory with the basic idea of algebraic quantum mechanics also in the case where the uncertainties of quantum states are specified as entropic uncertainities. That is to say, Definition 9.8, too, is consistent with algebraic quantum mechanics. This is reassuring because, as it was mentioned in Chapter 6, it is the Jordan algebra structure of a C^*-algebra only that bears physical meaning, thus one does not expect physical theories described by C^*-algebras having similar Jordan algebra structure to be different from the point of view of the uncertainty content they represent. This expectation is shown to be justified completely by Proposition 9.10. One can also express the message of Proposition 9.10 in a negative form by saying that if the entropic uncertainties of the quantum states on observables of an algebra are strictly smaller in every measurement situation than the entropic uncertainties of the quantum states on the corresponding observables of another algebra in the same measurement situation, then the Jordan algebra structures of the rescpective C^*-algebras must be regarded different.

The next natural question is whether Proposition 9.10 holds for entropic hidden theories i.e. whether $(\mathcal{N}, E(\mathcal{N}))$ can be an entropic hidden theory of a physical theory $(\mathcal{M}, E(\mathcal{M}))$ with respect to an $L \in (\mathcal{N}, \mathcal{M})$ conditional expectation. The general solution of this problem is not known; however, on the basis of the next negative result concerning the nonexistence of entropic hidden theory with respect to a conditional expectation in the finite dimensional case, it is conjectured here that the answer to this question is no.

Proposition 9.11 Let $\mathcal{M}_m, \mathcal{N}_n$ be full complex matrix algebras in finite dimensions m and n such that $n < m$. Then $(\mathcal{M}_m, E(\mathcal{M}_m))$ is not an entropic hidden theory of $(\mathcal{N}_n, E(\mathcal{N}_n))$ with respect to an $L \in D(\mathcal{M}_m, \mathcal{N}_n)$ conditional expectation.

Proof: In order that $\mathcal{N}_n$ be a subalgebra of $\mathcal{M}_m$ with a common unit, $m = nk$ must hold for some integer $k > 1$, and in this case $\mathcal{M}_m = \mathcal{N}_n \otimes \mathcal{M}_k$, where $\mathcal{N}_k$ is the full matrix algebra of complex $k \times k$ matrices, and then there exists a (unique) conditional expectation from $\mathcal{M}_m$ onto $\mathcal{N}_n$, which is given by

$$Lx = L(x_n \otimes x_k) = \text{Tr}(x_k)x_n \qquad x \in \mathcal{M}_m \qquad (9.25)$$

(Tr_k being the normalized trace over $\mathcal{N}_k$.) Let $A(n), B(n) \in \mathcal{N}_n$ and $A(k), B(k) \in \mathcal{N}_k$ be complementary operators in the respective algebras $\mathcal{N}_n$ and $\mathcal{N}_k$, i.e. $A(n), B(n); (A(k), B(k))$ are selfadjoint operators with non-degenerate spectra and eigenvectors $a(n)_i, b(n)_i$ $(i = 1 \ldots n)$ $(a(k)_i, b(k)_i$ $(i = 1, \ldots, k))$ such that

$$|\langle a(n)_i, b(n)_j \rangle| = \frac{1}{\sqrt{n}} \qquad i, j = 1, \ldots, n,$$

$$|\langle a(k)_i, b(k)_j \rangle| = \frac{1}{\sqrt{k}} \qquad i, j = 1, \ldots, k.$$

(see the Definition 2.4). As it was seen in Chapter 2 such complementary operators exist in every finite dimensional full matrix algebra; furthermore, if $A(n), B(n)$ and $A(k), B(k)$ are complementary, then the vectors $a(m)_{ij}$ and $b(m)_{ij}$ defined below and considered as eigenvectors of the operators

($A(m)$ and $B(m)$) respectively define a complementary observable pair ($A(m), B(m)$) in $\mathcal{M}_m$.

$$a(m)_{i,j} = a(n)_i \otimes a(k)_j \quad (i = 1, \ldots, n; j = 1, \ldots, k),$$
$$b(m)_{i,j} = b(n)_i \otimes b(k)_j \quad (i = 1, \ldots, n; j = 1, \ldots, k)$$

Assume now that $(\mathcal{M}_m, E(\mathcal{M}_m))$ is an entropic hidden theory of $(\mathcal{N}_n, E(\mathcal{N}_n))$ with respect to the conditional expectation given by (9.25). Then for an arbitrary state $\phi \in E(\mathcal{N}_n)$ there exists a μ such that (9.16)-(9.17) holds, which, for the complementary observables $A(m), B(m)$, yield the following inequalities

$$S_\phi(LA(m), \{d_i\}) \geq S_\omega(A(m), \{d_i\}) \tag{9.26}$$
$$S_\phi(LB(m), \{d_i\}) \geq S_\omega(B(m), \{d_i\}) \tag{9.27}$$

Since the eigenvalues do not play a role in the definition of complementarity, we may assume that all eigenvalues of the complementary operators are distinct, and we can take then a partition $\{d_i\}$ that isolates the eigenvalues. Then the entropies involved depend only on the spectral projectors and (dropping $\{d_i\}$) we denote them for short by $S_\phi(A(m))$ etc.. By (9.25) and by the special form of $A(m), B(m)$ we have $LA(m) = A(n)$ and $LB(m) = B(n)$, and so by adding (9.26) and (9.27) we arrive at

$$S_\phi(A(n)) + S_\phi(B(n)) \tag{9.28}$$
$$\geq S_\omega(A(m)) + S_\omega(B(m)) \tag{9.29}$$

The entropic uncertainty relation for finite dimensional complementary observables (Proposition 2.9) implies that the right hand side of (9.28) is greater than $\log(m)$ for every state ω. On the other hand, choosing ϕ to be one of the eigenstates of either $A(n)$ or $B(n)$ the left hand side becomes $\log(n)$, and so (9.28) is impossible since $m > n$. $\square$

9.4. The problem of local hidden variables

9.4.1. BELL'S QUESTION AND BELL'S INEQUALITY

Bell's work is generally considered a turning point in the history of the hidden variable problem for two reasons. One is that in Bell's formulation the hidden variable problem became a specific concistency question rather than a simple existence question. What Bell asked in his paper [11] was whether it is possible to eliminate in principle the statistical character of quantum mechanics without violating relativistic (locality) principles. The other reason why Bell's treatment of the problem is viewed as entirely novel is that his results (the Bell's inequalities) are interpreted as making consistency/inconsistency claims subject of empirical verification, since the inequalities in question can be empirically tested.

There are, in principle, two ways to investigate Bell's consistency question:

1. Given a hidden-variable theory T of quantum mechanics, one can try to formulate (consequences of) relativistic principles in T and see if this can be done consistently.
2. Given a form of quantum theory that does comply with relativistic principles explicitly, one can attempt to formulate the reduction of statistical character of this relativistic quantum theory.

Bell chose the option 1., and his treatment of the problem, put into mathematically more explicit terms than what one finds in his original papers, is the following.

Let us assume that the hypothetical hidden theory is a classical probability theory $(\Gamma, \Omega, \mathcal{E})$ with Ω being a σ-algebra of subsets of Γ and $\mathcal{E}$ being the set of probability measures on Ω, where the hidden variable character of the theory means that for every quantum observable Q there exist a function

$$f_Q : \Gamma \to \sigma_Q \subseteq \mathbb{R} \tag{9.30}$$

(σ_Q being the spectrum of the operator Q) such that for every quantum state ϕ there exists a μ measure in $\mathcal{E}$ that yields the quantum expectation values in the form of an integral:

$$\langle Q \rangle_\phi = \int_\Gamma f_Q \mathrm{d}\mu \tag{9.31}$$

The elements of Γ are the hypothetical "hidden variables", and $f_Q(\lambda) \in \sigma_Q$ is interpreted as the "value of the observable Q, if the value of the hidden parameter is equal to λ"; this is the value one finds in a measurement of the observable Q.

Assume now that we are given a joint quantum system S consisting of two parts S_1 and S_2 that are located in two disjoint regions of space that are far apart. Let Q_1, Q_2 and R_1, R_2 be four bounded quantum observables of the joint system such that Q_1, Q_2 are non-commuting observables of system S_1, the R_1, R_2 are non-commuting observables of system S_2, and such that $\| Q_i \| \leq 1$ and $\| R_i \| \leq 1$ ($i = 1, 2$). We may form then the observables

$$Q_1 R_1 \quad Q_1 R_2 \quad Q_2 R_1 \quad Q_2 R_2$$

If we assume the existence of a hidden theory in the above sense for the system S, then there exist the functions

$$f_{Q_1} \quad f_{R_1} \quad\quad f_{Q_2} \quad f_{R_2}$$

$$f_{Q_1R_1} \quad f_{Q_1R_2} \quad f_{Q_2R_1} \quad f_{Q_2R_2}$$

representing the observables indicated, and we may then compute the following expectation value

$$\begin{aligned}\langle Q_1R_1 + Q_1R_2 + Q_2R_1 - Q_2R_2\rangle_\phi &= \phi(Q_1(R_1+R_2) + Q_2(R_1-R_2)) \\ &= \int_\Gamma (f_{Q_1R_1} + f_{Q_1R_2} + f_{Q_2R_1} - f_{Q_2R_2})\mathrm{d}\mu\end{aligned}$$

If we assume that

$$f_{Q_iR_j}(\lambda) = f_{Q_i}(\lambda)f_{R_j}(\lambda) \qquad \lambda \in \Gamma \quad (i,j=1,2) \tag{9.32}$$

then we have

Proposition 9.12

$$-1 \leq \int_\Gamma (f_{Q_1R_1} + f_{Q_1R_2} + f_{Q_2R_1} - f_{Q_2R_2})\mathrm{d}\mu \leq 1 \tag{9.33}$$

(We shall prove this proposition under more general conditions in Chapter 10.)

The inequality in the above Proposition (9.12) is called *Bell's inequality*, and the interpretation of eq. (9.32) is that it is the expression of "physical locality" in terms of (on the level of the) the hidden variable theory.

Proposition 9.12 says that the hidden variable theory predicts the bound 1 for the correlation

$$|\langle Q_1R_1 + Q_1R_2 + Q_2R_1 - Q_2R_2\rangle_\phi| \tag{9.34}$$

Consequently, if the bound is found to be violated in experiment, some assumptions that go into the derivation of Bell's inequality are empirically refuted. The bound 1 has been checked experimentally, and the general interpretation of the results is that the inequality (9.33) is violated.

The significance of this result together with the question of what precisely is the content of the notion of physical locality that is supposed to be expressed by the condition (9.32) have been subjects of intense debates and controversies. These we do not wish to review here, mainly because in the next section we take the other approach to Bell's consistency question indicated as option 2.: We wish to formulate the reduction of the statistical character of a manifestly *local and relativistic* quantum mechanics, namely relativistic quantum field theory.

Before embarking on this, we wish to mention a closely related definition of hidden variable theory, put forward – and shown to be impossible – by Kochen and Specker.

Kochen and Specker [85] define $(\Gamma, \Omega, \mathcal{E})$ to be a hidden variable theory of quantum mechanics if the three conditions below are satisfied.

THE PROBLEM OF HIDDEN VARIABLES

1. For every quantum observable (selfadjoint operator) Q there exists a function $f: \Gamma \to \mathbb{R}$,
2. For every quantum state ϕ there exists a probability measure $\mu \in \mathcal{E}$ such that

$$\langle Q \rangle_\phi = \int_\Gamma f_Q \, d\mu \qquad (9.35)$$

3. If $g: \mathbb{R} \to \mathbb{R}$ is a function such that $g(Q)$ is a selfadjoint operator – (defined via the spectral theorem, see (2.6)) – then the hidden variable representative $f_{g(Q)}$ of $g(Q)$ is just $g \circ f$, i.e. we have

$$f_{g(Q)}(\lambda) = g(f(\lambda)) \qquad \lambda \in \Gamma \qquad (9.36)$$

This last condition requires that the algebraic structure of the observables be mirrored in a way in the hidden variable representation. The motivation behind demanding this condition 3. is the following reasoning. If we assume that the measurement of any observable Q reveals the values $f_Q(\lambda)$ the quantity Q has in the hidden states λ, then the quantum mechanical definition of $g(Q)$ implies that the statistical distribution of the values $g(f_Q(\lambda))$ must be identical with the quantum mechanical statistical distribution of the values of the observable $g(Q)$ (provided $g(Q)$ exists), and this latter distribution must be identical with the statistical distribution of values $f_{g(Q)}$ – otherwise the hidden theory could not reproduce the quantum mechanical expectation value of $g(Q)$ by eq. (9.35).

Now it is easily seen that the assumption of a hidden variable theory of quantum mechanics that satisfies (9.36) implies the existence of a partial algebra homomorphism h from the partial algebra $\mathcal{B}(\mathcal{H})$ into the commutative algebra of real valued functions defined on Γ. Indeed, let $Q_1, Q_2 \in \mathcal{B}(\mathcal{H})$ be two arbitrary, commuting selfadjoint operators. Then there exists a selfadjoint operator Q such that

$$Q_1 = g_1(Q) \quad \text{and} \quad Q_2 = g_2(Q)$$

and we have

$$r_1 Q_1 + r_2 Q_2 = (r_1 g_1 + r_2 g_2)(Q) \qquad (9.37)$$
$$Q_1 Q_2 = (g_1 g_2)(Q) \qquad (9.38)$$

for any real numbers r_1, r_2. Hence if we define h by

$$h(Q) = f_Q \qquad (9.39)$$

then using (9.36) and (9.37) we obtain

$$h(r_1 Q_1 + r_2 Q_2)(\lambda) = h((r_1 g_1 + r_2 g_2)(Q))(\lambda)$$
$$= f_{(r_1 g_1 + r_2 g_2)(Q)}(\lambda) = (r_1 g_1 + r_2 g_2)((f_Q(\lambda))$$
$$= r_1 g_1(f_Q(\lambda)) + r_2 g_2(f_Q(\lambda)) = r_1 f_{g_1(Q)}(\lambda) + r_2 f_{g_2(Q)}(\lambda)$$
$$= r_1 h(g_1(Q))(\lambda) + r_2 h(g_2(Q))(\lambda) = r_1 (hQ_1)(\lambda) + r_2 (hQ_2)(\lambda)$$

and similarly

$$\begin{aligned}
h(Q_1Q_2)(\lambda) &= h((g_1g_2)(Q))(\lambda) \\
= f_{g_1g_2(Q)}(\lambda) &= g_1(f_Q(\lambda))g_2(f_Q(\lambda)) \\
= f_{g_1(Q)}(\lambda)f_{g_2(Q)}(\lambda) &= h(Q_1)h(Q_2)(\lambda)
\end{aligned}$$

which means that h is a partial algebra homomosphism. Since there exists no partial algebra homomorphism from $\mathcal{B}(\mathcal{H})$ into a commutative algebra if $\dim(\mathcal{H}) \geq 3$ (Proposition 4.10), there exist no hidden variable theory of quantum mechanics in the sense of Kochen and Specker.

9.4.2. NO-GO PROPOSITION ON DISPERSIVE LOCAL HIDDEN THEORIES

In this subsection we wish to investigate Bell's consistency question by adopting the strategy of trying to formulate the reduction of the statistical character of a manifestly local and relativistic quantum mechanics (namely relativistic local quantum field theory), retaining at the same time the local relativistic structure. By a local relativistic quantum mechanics is meant here the algebraic quantum field theory as described in Chapter 10.1, for which the reduction of the statistical character in the sense of Definition 9.2 makes perfect sense; so the task now is to specify what is meant by "preserving the local relativistic structure" in the reduction.

Recall that the reduction of the statistical content of the physical theory $(\mathcal{A}, E(\mathcal{A}))$ by a hidden theory $(\mathcal{B}, E(\mathcal{B}))$ with respect to a positive map $L \in \mathcal{D}(\mathcal{B}, \mathcal{A})$ means that for every state $\phi \in E(\mathcal{A})$ the state $\phi \circ L \in E(\mathcal{B})$ can be decomposed in the form

$$\phi(LB) = \int \omega(B)\mathrm{d}\mu(\omega) \qquad B \in \mathcal{B} \tag{9.40}$$

in such a way that the dispersions of the states ω that are in the support of the decomposing measure μ are strictly smaller than the dispersion of ϕ (see the Definition 9.2 for a precise formulation). If $(\mathcal{A}, E(\mathcal{A}))$ and $(\mathcal{B}, E(\mathcal{B}))$ are both physical theories with $\mathcal{A}$ and $\mathcal{B}$ being quasilocal C^*-algebras in the sense of ARQFT, then a natural way to define what it means to preserve the local relativistic structure during the reduction of statistical content of $(\mathcal{A}, E(\mathcal{A}))$ is to demand that both L and μ in eq. (9.40) respect the local relativistic structure of the local algebras involved.

So let $\mathcal{A}$ and $\mathcal{B}$ be two relativistic quasilocal algebras, and denote by β the representation of the Poincaré group $\mathcal{P}$ on $\mathcal{B}$. We wish to define local properties of the positive, unit preserving map $L: \mathcal{B} \to \mathcal{A}$, by which we mean properties that express the similarity of the relativistic quasilocal structure of $\mathcal{A}$ and $\mathcal{B}$. In the following definition (i) is a natural locality demand, the

THE PROBLEM OF HIDDEN VARIABLES 165

content of (ii) is that L does not destroy the relativistic covariance, whereas (iii) is of technical nature.

Definition 9.9 The positive, unit preserving map $L: \mathcal{B} \to \mathcal{A}$ between two relativistic quasilocal algebras $\mathcal{A}$ and $\mathcal{B}$ is called *Einstein local* if
(i) L maps the local algebras into local algebras,
(ii) L commutes with the two representations α and β in the sense that

$$\alpha_g \circ L = L \circ \beta_g \quad g \in \mathcal{P} \tag{9.41}$$

(iii) the restriction of L to each local algebra is continuous in the ultraweak operator topology.

To motivate the property μ should possess, let us first define the restriction maps $r_V: E(\mathcal{B}) \to E(\mathcal{B}(V))$ by

$$(r_V \phi)(X) \equiv \phi(X) \qquad X \in \mathcal{B}(V) \tag{9.42}$$

This r_V is continuous if the state spaces are taken with their respective w^*-topology, hence if $E_V \subseteq E(\mathcal{B}(V))$ is a Borel subset of local states, then $r^{-1}(E_V) \subseteq E(\mathcal{B})$ is the Borel set of states on the quasilocal algebra that extend the local states in E_V. Let V_1 and V_2 be two spacelike separated regions, and

$$E_{V_1} \subseteq E(\mathcal{B}(V_1)) \qquad E_{V_2} \subseteq E(\mathcal{B}(V_2)) \tag{9.43}$$

Then the Borel subset $G^1{}_2$ defined by

$$G^1{}_2 \equiv r^{-1}{}_{V_1}(E_{V_1}) \setminus r^{-1}{}_{V_2}(E_{V_2}) \tag{9.44}$$

consists of states over the quasilocal algebra $\mathcal{B}$ that extend the states in E_{V_1} without extending any of the states in E_{V_2}. This means that no information about the local algebra $\mathcal{B}(V_2)$ contained in E_{V_2} can be obtained from $G^1{}_2$, and, similarly, the states in the accordingly defined $G^2{}_1$ do not contain information about the local algebra $\mathcal{B}(V_1)$. It is a natural idea then to implement physical locality on the level of the averaging process specified by eq. (9.40) by demanding that μ "does not mix up" the states in $G^1{}_2$ and $G^2{}_1$, in other words, the two functionals obtained by averaging via μ over $G^1{}_2$ and $G^2{}_1$ respectively

$$\phi_1 \equiv \int_{G^1{}_2} \omega d\mu(\omega) \tag{9.45}$$

$$\phi_2 \equiv \int_{G^2{}_1} \omega d\mu(\omega) \tag{9.46}$$

should be "independent".

A natural independence condition of positive linear functionals over C^*-algebras is their disjointness: Two positive functionals ϕ_1 and ϕ_2 are defined to be disjoint, if the two representations π_1 and π_2 determined by ϕ_1 and ϕ_2 via the GNS construction (Section 6.1) are disjoint, i.e. they do not have unitary equivalent subrepresentations. Let us call a μ on $E(\mathcal{B})$ *local* if for any two Borel sets E_1, E_2 the functionals

$$\phi_1 = \int_{E_1} \omega d\mu(\omega) \quad \text{and} \quad \phi_2 = \int_{E_2} \omega d\mu(\omega)$$

are disjoint. This condition can be shown to be equivalent to the condition that for any Borel set $E \subseteq E(\mathcal{B})$ the two functionals

$$\phi_E = \int_E \omega d\mu(\omega) \quad \text{and} \quad \phi_{E(\mathcal{B})\setminus E} = \int_{E(\mathcal{B})\setminus E} \omega d\mu(\omega)$$

are disjoint. A μ measure having this property is called a *subcentral measure*.

We can now give the definition of Einstein local hidden theory.

Definition 9.10 Let $\mathcal{A}$ and $\mathcal{B}$ be two relativistic quasilocal algebras in the sense of ARQFT. $(\mathcal{B}, E(\mathcal{B}))$ is said to be a local hidden theory of $(\mathcal{A}, E(\mathcal{A}))$ if

(i) L is Einstein local in the sense of Definition 9.9,
(ii) $(\mathcal{B}, E(\mathcal{B}))$ is a hidden theory of $(\mathcal{A}, E(\mathcal{A}))$ via L in the sense of Definition 9.2, and
(iii) the measure μ that decomposes the states $L^*\phi$ in the sense of (9.40) can be chosen subcentral.

A proposition stating the nonexistence of a local hidden theory for a given covariant net of local algebras shows then what locality properties cannot be preserved during a hypothetical reduction of the statistical content of the local theory – where the statistical content is measured in terms of the dispersions of the states.

Before formulating a negative proposition on the existence of local hidden theory, let us recall a few definitions and facts that will be used in the proof. A state ϕ on $\mathcal{A}$ is called *locally normal* if the restriction of ϕ to every local von Neumann algebra $\mathcal{N}(V)$ is ultraweakly continuous. The state ϕ is said to be *locally faithful* if the condition $x > 0$ implies $\phi(x) > 0$ for any strictly local element $x \in \mathcal{N}(V)$.

Let μ be a subcentral measure on $E(\mathcal{B})$ that decomposes a state $\psi \in E(\mathcal{B})$ in the sense of (9.40). Then if ψ is a factor state, i.e. if the center $Z_\psi = \pi_\psi(\mathcal{B})'' \cap \pi_\psi(B)'$ of the von Neumann algebra $\pi_\psi(\mathcal{B})''$ generated by ψ in the Gel'fand-Naimark-Segal (GNS) representation π_ψ consists of the multiples of the identity only, then μ is the Dirac measure δ_ψ concentrated at ψ.

THE PROBLEM OF HIDDEN VARIABLES   167

Proposition 9.13 *Let $(\mathcal{A}, E(\mathcal{A}))$ and $(\mathcal{B}, E(\mathcal{B}))$ be two local relativistic quantum field theories satisfying the standard axioms. If there is a locally normal, locally faithful, α-invariant, translation clustering state ϕ on $\mathcal{A}$ which is not dispersion-free, then $(\mathcal{A}, E(\mathcal{A}))$ does not have a local hidden theory via L in the sense of Definition 9.10.*

Proof: One shows first that if there is a state $\psi \in E(\mathcal{A})$ with nonzero dispersion $\sigma_\psi(Lx_0) > 0$ on some selfadjoint $x_0 \in \mathcal{B}$, then $(\mathcal{B}, E(\mathcal{B}))$ cannot be a hidden theory of $(\mathcal{A}, E(\mathcal{A}))$ via L, if the only measure that decomposes $L^*\psi \in E(\mathcal{B})$ is the Dirac measure $\delta_{L^*\psi}$. To show this assume that $(\mathcal{B}, E(\mathcal{B}))$ is a hidden theory of $(\mathcal{A}, E(\mathcal{A}))$ via L. Proceeding just like in the proof of Proposition 9.5 we obtain

$$\psi((Lx_0)^2 - Lx_0^2) > \psi(Lx_0)^2 - \int \omega(x_0)\mathrm{d}\mu(\omega) = 0 \qquad (9.47)$$

which is a contradiction since the left hand side of (9.47) is not greater than zero by the Cauchy-Schwartz inequality (eq. (9.2)) for L.

Thus to prove the proposition it is enough to show that there is a non-dispersion-free state $\psi \in E(\mathcal{A})$ such that $L^*\psi \in E(\mathcal{B})$ is a factor state. We prove that $L^*\phi$ is a factor state over $\mathcal{B}$ by showing that the assumption of $L^*\phi$ not being a factor state contradicts the clustering property of ϕ.

Since ϕ is α-invariant, $L^*\phi$ is β-invariant by (9.41) of definition of Einstein locality of L. Therefore both α and β are implemented by unitary representations U and W in the GNS representations π_ϕ and $\pi_{L^*\phi}$ of $\mathcal{A}$ and $\mathcal{B}$, respectively. Let L_π denote the "representation" of L in π_ϕ and $\pi_{L^*\phi}$, i.e. $L_\pi(\pi_{L^*\phi}(x)) = \pi_\phi(Lx)$. Then (9.41) in Definition 9.9 takes the form

$$U_g L_\pi(\bullet) U_g^* = W_g L_\pi(\bullet) W_g^* \qquad (9.48)$$

Assume that $L^*\phi$ is not a factor state. Then there is a non-trivial projector $\pi_{L^*\phi}(P)$ in the center

$$\begin{aligned} Z &= \pi_{L^*\phi}(\mathcal{B})'' \cap \pi_{L^*\phi}(\mathcal{B})' \\ &= \cap_{V \subset M}(\pi_{L^*\phi}(\mathcal{B}(V)))'' \cap \pi_{L^*\phi}(\mathcal{B}(V))' \end{aligned}$$

(The last iquality follows because $\mathcal{B}$ is a quasilocal algebra.) By the local normality of ϕ and L, $L^*\phi$ is locally normal too, and so $\pi_{L^*\phi}(\mathcal{B}(V))'' = \pi_{L^*\phi}(\mathcal{B}(V))$. Thus $\pi_{L^*\phi}(P)$ is a nontrivial projector contained in (the center of) each local algebra $\pi_{L^*\phi}(\mathcal{B}(V))$. Fix a V. Now L maps the local algebras into local ones, thus $L_\pi \pi_{L^*\phi}(P)$ is an element in some local algebra $\pi_\phi \mathcal{A}(V')$, say. We may assume that $L_\pi \pi_{L^*\phi}(P)$ is nonzero for if it were, then we could take the orthogonal complement $\pi_{L^*\phi}(P)^\perp$, which shares all the properties of $\pi_{L^*\phi}(P)$ stated so far, and the unit preserving property of L implies that $L_\pi\pi_{L^*\phi}(P)$ and $L_\pi\pi_{L^*\phi}(P)^\perp$ cannot both be zero. So $L_\pi\pi_{L^*\phi}(P)$ is a nonzero element, which is also positive by positivity of L, and this implies that $L_\pi\pi_{L^*\phi}(P)$ has a nontrivial spectral projector $L_\pi\pi_{L^*\phi}(e)$. By the theorem of Araki (Proposition 10.4) the elements of Z commute with W_g ($g \in \mathcal{P}_T$), $\mathcal{P}_T$ being the translation subgroup of $\mathcal{P}$, and so $L_\pi\pi_{L^*\phi}(P)$ commutes with U_g ($g \in \mathcal{P}_T$) by (9.48). But then U_g commutes with $L_\pi\pi_{L^*\phi}(e)$, too, and it follows that $L_\pi\pi_{L^*\phi}(e)$ is a $U(g)$-invariant nontrivial projector in $\pi_\phi\mathcal{A}(V')$. Let $\pi_\phi(R) \in \pi_\phi\mathcal{A}(V')$ be any nonzero local projector orthogonal to $L_\pi\pi_{L^*\phi}(e)$. Since ϕ is locally faithful we have

$$\langle \Omega_\phi, \pi_\phi(R)\Omega_\phi \rangle > 0 \quad \text{and} \quad \langle \Omega_\phi, L_\pi\pi_{L^*\phi}(e)\Omega_\phi \rangle > 0 \qquad (9.49)$$

where Ω_ϕ is the cyclic vector representing ϕ in the π_ϕ representation. On the other hand, by the clustering property of ϕ and by the $U(g)$-invariance of $L_\pi\pi L^*\phi(e)$ the equation

$$\begin{aligned}
0 &= \langle \Omega_\phi, \pi_\phi(R) L_\pi\pi_{L^*\phi}(e)\Omega_\phi \rangle \\
&= \lim_{t\to\infty} \langle \Omega_\phi, \pi_\phi(R) U_{tg} L_\pi\pi_{L^*\phi}(e) U^*_{tg}\Omega_\phi \rangle \\
&= \langle \Omega_\phi, \pi_\phi(R)\Omega_\phi \rangle \langle \Omega_\phi L_\pi\pi_{L^*\phi}(e)\Omega_\phi \rangle
\end{aligned}$$

must hold too, which contradicts (9.49). □

Let ϕ be a vacuum state in $E(\mathcal{A})$. Then by the Reeh-Schlieder theorem (Proposition 10.1) ϕ is cyclic and separating for the local algebras, which means that in this case ϕ is locally faithful too. Obviously, the requirement that the vacuum state is not dispersion-free, is not a strong one, moreover, if the vacuum state ϕ in $E(\mathcal{A})$ is the unique α-invariant state, then ϕ is known to have also the translation clustering property (both in massive and in massless theories). In this case, by (ii) of Einstein locality of L, $L^*\phi$ is the unique β-invariant state, therefore by the assumption that $E(\mathcal{B})$ contains at least one vacuum state, $L^*\phi$ is the (unique) vacuum state in $E(\mathcal{B})$; in particular, the spectrum condition is fulfilled in $\pi_{L^*\phi}$. So if $(\mathcal{A}, E(\mathcal{A}))$ is such that the vacuum state ϕ is a unique α-invariant, locally normal state then the assumptions of Proposition 9.13) are fulfilled, and the above proposition tells us that such $(\mathcal{A}, E(\mathcal{A}))$ relativistic quantum field theories are the best possible ones, if "best" means "containing the least statistical uncertainty", i.e., the statistical uncertainty inherent in the description of quantum fields by these theories cannot be reduced without violating at least one of the Einstein local properties as speficied in Definition 9.10.

Note that "statistical uncertainty" has been assumed in this section being measured by the dispersion of quantum states. However, as we have seen in Section 9.3, the dispersion is not the only conceivable measure of uncertainty of a probability distribution that one can utilize in the definition of a hidden variable theory: the entropy also expresses a kind of uncertainty, which is conceptually different from what is expressed by dispersion. So one could define the notion of entropic Einstein local hidden theory along the lines of the present section, and one could raise the problem of existence of entropic local hidden theory. Since no results are known in this direction, we do not give the definitions in question.

9.5. Bibliographic notes

There is an enormous literature on the hidden variable problem. A comprehensive review of the hidden variable problem up to the early seventies can be found in [9] and in [73]. The algebraic approach to the hidden variable problem, especially the Definition 9.2, and the propositions 9.5, 9.6, 9.8 and 9.9 on the nonexistence of dispersive hidden theories are taken from [120]. The approximate hidden variable problem is defined in algebraic terms in [119]. The idea of entropic hidden theories was proposed in [122], Propositions 9.10 and 9.11 are from [122] and [124]. The notion of local (dispersive) hidden theory was first defined in terms of quasilocal algebras describing non-relativistic local quantum field theory, see [121], the relativistic version and Proposition 9.13 is from [123]. The theory of quasilocal C^*-algebras is described e.g. in Chapter 2.6 in [28]. The Kochen-Specker notion of hidden variables was proposed in [85], where an explicit 117 element finite sublattice of $\mathcal{P}(\mathcal{H})$ is constructed on which there exists no evaluation map (i.e. a partial Boolean algebra homomorphism). There are much smaller sublattices of $\mathcal{P}(\mathcal{H})$ on which an evaluation map cannot exist, see [41]. The question of how to select a sublattice of $\mathcal{P}(\mathcal{H})$ elements of which could be viewed as having simultaneuosly definite truth values is investigated in [14], where the notion of quasiBoolean lattice is introduced. For a philosophical analysis of the Kochen-Specker notion see Chapters 5 and 6 in [134].

CHAPTER 10

Violation of Bell's inequality in quantum field theory

The aim of this chapter is to review some results obtained in the past decade on the violation of Bell's inequality in algebraic relativistic quantum field theory (ARQFT). We recall first the basic idea and the standard axioms of ARQFT in Section 10.1, this is followed by recalling a few important definitions and theorems that characterize ARQFT. The presentation will be restricted to those structure elements, definitions and theorems of ARQFT that are necessary to comprehend the statements on the violation of Bell's inequality, or else are referred to explicitly somewehere in the book. The results in Section 10.1 are stated without proofs, which in most cases are very technical, and giving them would lead us too far away from the central theme of this work. Section 10.2 introduces the notion of Bell correlation for two C^*-algebras $\mathcal{A}$ and $\mathcal{B}$ in a state ϕ (Definition 10.8), and it is proved (Proposition 10.11) that in the category of C^*-algebras the value $\sqrt{2}$ is the upper bound for the Bell correlation. Proposition 10.10 gives sufficient conditions on ϕ, $\mathcal{A}$ and $\mathcal{B}$ implying that the Bell correlation is bounded by 1, which is Bell's inequality. It is proved then (Proposition 10.12) that if $\mathcal{A}$ and $\mathcal{B}$ are local observable algebras in ARQFT, possibly pertaining to spacelike separated spacetime regions, then there exist states in which the Bell correlation for the algebras takes on its maximal value ($\sqrt{2}$), i.e. Bell's inequality is maximally violated in ARQFT. The main purpose of the review of these results is to formulate clearly and precisely this consequence of violation of Bells inequality in ARQFT: relativistic quantum field theory predicts superluminal probabilistic correlations, i.e. correlations between quantum event pairs that belong to spacelike separated spacetime regions. The chapter is closed with pointing out the presence in ARQFT of these superluminal correlations, and two possible strategies of explanation of the correlations are then formulated in Section 10.4. One strategy is to claim that there is a direct causal connection between the correlated events; another is trying to find a common probabilistic cause of the correlations. These two options will be explored in the subsequent two chapters 11 and 12.

10.1. Basic notions of algebraic quantum field theory

The bounded observables in ordinary Hilbert space quantum mechanics are represented by selfadjoint elements of a C^*-algebra. But the observables in ordinary quantum mechanics are not viewed as carrying a label, index or parameter that would show where these observables are located in space or in time: It would not make sense to ask "where" and at "what time" the spin observable "is"; nor does the formalism contain means to express that a certain preparation of a state (i.e. a projection) "takes place" in a certain region of space at a certain time. In short, ordinary quantum mechanics is a non-local theory. Nor is it Lorentz covariant: it has the Galilean group as its symmetry group, and the observables are identified as the generators of the (continuous) projective representation of this group.

These two features of standard quantum mechanics are unsatisfactory, and the intuitive idea of local algebraic relativistic quantum field theory (ARQFT) is that the operational physical notions such as observable, preparation, measurement, etc. are meaningful only if they are local in the sense that they are considered as pertaining to a bounded region V in the Minkowski spacetime M. Further, a *relativistic* quantum theory also should be Lorentz covariant.

The mathematical implementation of the idea of local, relativistic algebraic quantum field theory is then that the basic object in the mathematical model of a quantum field is the association of a C^*-algebra $\mathcal{A}(V)$ to regions V in the Minkowski spacetime M, and that all the physical content of the theory should be contained in the assignment $V \mapsto \mathcal{A}(V)$. In particular, Lorentz covariance of the theory should also be expressed in terms of the *net of algebras* $(\{A(V)\}, V \subset M)$, the net of algebras of local observables.

Thus in ARQFT one postulates certain properties of the net of local observable akgebras $(\{A(V)\}, V \subset M)$, properties that are dictated by physical considerations. Below we list these postulates.

Definition 10.1 The assignment

$$V \mapsto \mathcal{A}(V) \qquad V \subset M \quad \text{open, bounded} \tag{10.1}$$

is called a *covariant net of strictly local observables* if the following (i)-(iv) conditions are satisfied.

(i) isotony: $\mathcal{A}(V_1) \subset \mathcal{A}(V_2)$ if $V_1 \subset V_2$
(ii) microcausality: $\mathcal{A}(V_1)$ commutes with $\mathcal{A}(V_2)$ if V_1 and V_2 are spacelike separated.

Let

$$\mathcal{A}_0 \equiv \cup_V \mathcal{A}(V)$$

then $\mathcal{A}_0$ is a normed $*$-algebra, completion of which is a C^*-algebra, called the *quasilocal algebra* determined by the net ($\{A(V)\}, V \subset M$).

(iii) **relativistic covariance**: there exists a representation α of the Poincaré group $\mathcal{P}$ by automorphisms $\alpha(g)$ on $\mathcal{A}$ such that

$$\alpha(g)\mathcal{A}(V) = \mathcal{A}(gV) \qquad (10.2)$$

for every $g \in \mathcal{P}$ and for every V.

It also is postulated that there exists at least one physical representation of the algebra $\mathcal{A}$, that is to say, it is required that there exist a Poincaré invariant state ϕ_0 (vacuum) such that the spectrum condition ((iv) below) is fulfilled in the corresponding cyclic (GNS) representation (see Section 6.1) ($\mathcal{H}, \Omega, \pi$). In this representation of the quasilocal algebra the representation α is given as a unitary representation, and there exist the generators P_i, ($i = 0, 1, 2, 3$) of the translation subgroup of the Poincaré group $\mathcal{P}$. The spectrum condition is formulated in terms of these generators as

(iv) **spectrum condition**:

$$P_0^2 \geq 0, \qquad P_0^2 - P_1^2 - P_2^2 - P_3^2 \geq 0 \qquad (10.3)$$

To formulate axiom (v) we need the definition of *causal hull* (or *causal shadow*) V^- of a spacetime region V: V^- is defined to be the set of points x in the Minkowski spacetime that have the property that *every* timelike straight line containing x intersects V.

(v) **local primitive causality**: If V^- denotes the causal hull of region V, then

$$\mathcal{A}(V^-) = \mathcal{A}(V)$$

Remarks on the postulates:

1. Condition (i) is a natural consistency postulate: it expresses that what is measurable in a region V_1 also should be measurable in a larger region V_2 containing V_1.
2. Condition (ii) is an expression of the intuition, rooted in the special theory of relativity, that measurements of observables that are located in spacelike separated spacetime regions should not disturb each other, they should be copossible. To illustrate this point assume that the observable $B \in \mathcal{A}(V_1)$ has a purely discrete spectrum with spectral resolution $B = \sum_i \lambda_i P_i$. Then $\sum_i \phi(P_i A P_i)$ is the expectation of the observable $A \in \mathcal{A}(V_2)$ in the state obtained after preparing ϕ and subsequently measuring B. Since A and B commute, we have $\sum_i \phi(P_i A P_i) = \phi(A)$, that is the expectation of A in ϕ is independent of whether a (non-selective) measurement of B is carried out between the preparation of ϕ and measuring A.

3. The spectrum condition is the requirement of positivity of energy.
4. Local primitive causality is an expression of the hyperbolic character of the time evolution in ARQFT. Local primitive causality says that what is observable in a region V in the Minkowski spacetime, is determined by what is observable in a region that causally determines the region V.

Given a state (e.g. the vacuum) one can consider the net in the corresponding cyclic π representation and, identifying the weak closures $\pi(\mathcal{A}(V))''$ of the local C^*-algebras with the observable algebras in the state ϕ_0, one can assume that in the representation given by a state the net consists of local von Neumann algebras $\pi(\mathcal{A}(V))'' = \mathcal{N}(V)$ for which the condtions (i)-(iv) hold. In what follows we always assume that a net of von Neumann algebras is in fact a representation of a covariant net of C^*-algebras in a vacuum representation with a unique vacuum state.

Algebraic quantum field theories given by local nets $(\mathcal{A}, \mathcal{A}(V))$ satisfying the postulates (i)-(v) only are very general, and the net typically has some further properties implied by the particular features of the nets that can be constructed. Another way to go beyond the axioms (i)-(v) is by supplementing them with some further, physically motivated constraints on the net. This is necessary if one wants to be able to prove (even rather general) theorems on the field theory given by the net (such as Reeh-Schlieder theorem, types of local algebras etc.). Such an additional property is weak additivity:

(vi) Let $\mathcal{N}(V)$ be a net of von Neumann algebras on a Hilbert space $\mathcal{H}$. We say that *weak additivity* holds for $\mathcal{N}(V)$ if for any (possibly unbounded) region O in M we have

$$\mathcal{N}(O) = \{\mathcal{N}(V) \mid V \subset O\}'' \qquad (10.4)$$

Typical unbounded regions are the wedge regions. The *right wedge* W_R is defined by

$$W_R \equiv \{x \in M \mid x > |x_0|\} \qquad (10.5)$$

and the *left wedge* W_L by

$$W_L \equiv \{x \in M \mid x_1 < |x_0|\} \qquad (10.6)$$

The set of all wedge regions is the set of all Poincaré transforms of W_R.

If for an arbitrary spacetime region $V \subset M$ the *causal complement* V' of V is defined by

$$V' \equiv \{x \in M \mid x \text{ is spacelike separated from every element in } V\} \qquad (10.7)$$

then $W'_R = W_L$. Two such wedge regions are called *complementary*. Complementary wedges are *tangent* regions: their closures intersect at one point.

The interpretation of the weak additivity condition (vi) is that the spacetime is homogeneous, there does not exist "minimal distance", and in particular, (v) implies that the wedge algebras are locally generated:

$$\mathcal{N}(W) = \{\mathcal{N}(V) \mid V \subset W\}'' \tag{10.8}$$

An additional reason why assuming (vi) is important is that it is an assumption needed in one of the most important theorems in ARQFT, the Reeh-Schlieder Theorem:

Proposition 10.1 (Reeh-Schlieder Theorem) *Let $\mathcal{N}(V)$ be a net of local von Neumann algebras arising as the vacuum representation of a covariant net of local C^*-algebras satisfying the standard axioms (i)-(iv), assume furthermore that the net $\mathcal{N}(V)$ also has the weak additivity property. Then the vacuum vector is both cyclic and separating for any local algebra $\mathcal{N}(V)$ such that the causal complement V' of V is non-empty. Thus, in this case the vacuum state ϕ_0 is faithful on local algebras whose causal complement is non-trivial.*

Another important class of spacetime regions is the set of *double cones*. Let $y \in M$ lie in the forward light cone of $x \in M$. Then the nonempty interior of the intersection of x's forward light cone with y's backward light cone is called a double cone K. The double cones obtained in this manner for all x, y form the set $\mathcal{K}$ of all double cones. Two double cones are said to be *tangent* if they are spacelike separated and their closures intersect at one single point.

We mention two further possible properties of a net. The first of these, called *wedge duality*, is a strengthening of the locality axiom, the second one, called *cone cyclicity of vacuum* requires that in an irreducible vacuum representation the cone algebras already generate the representation space. That is to say, already local operations are sufficient to determine the representation space.

$$\mathcal{A}(W') = \mathcal{A}(W)' \qquad \text{(wedge duality)} \tag{10.9}$$

Ω is cyclic for $\cup_K \mathcal{A}(K)$ \qquad (cone cyclicity of vacuum) \qquad (10.10)

Both wedge duality and the cone cyclicity of the vacuum are satisfied if the net is associated to a Wightman field in a certain precise sense. Roughly, the association of a net of local algebras to a Wightman field Φ means that the local algebras $\mathcal{A}(V)$ are generated by the field operators smeared with test functions having support in the spacetime region V. "Being generated" means that the smeared field operators $\Phi[f]$ with $\operatorname{supp} f \subset V$ have selfadjoint extensions affiliated with the von Neumann algebra $\mathcal{N}(V)$.

Intuitively the relation of a net to quantum fields (in the sense of Wightman axioms) resembles the relation between a manifold and a coordinate system: The nets are the more fundamental objects, which make sense and can be defined without reference to fields, and different fields may describe the same local net.

Definition 10.2 A covariant net $\{\mathcal{N}(V)\}$ of local von Neumann algebras on the Hilbert space $\mathcal{H}$ is called *dilatation-invariant* if there exist a strongly continuous unitary representation D on $\mathcal{H}$ of the dilatation group on the Minkoski space M such that

$$D(\lambda)\mathcal{N}(V)D(\lambda)^{-1} = \mathcal{N}(\lambda V)$$
$$D(\lambda)\Omega = \Omega$$

where Ω is the unique vacuum vector, and where

$$\lambda V \equiv \{\lambda x \mid x \in V\}$$

for any $0 < \lambda \in \mathbb{R}$ is the dilatation group.

A characteristic property of algebraic field theory that distinguishes it from ordinary, non-local, non-relativistic quantum mechanics is that the local algebras are very different from the observable algebra of the standard quantum mechanics: the local algebras are typically infinite, and for specific spacetime regions they can be proven to be of type **III** (see the classification theory of von Neumann algebras in Chapter 6, especially Section 6.2). More precisely we have the following two results, which spell out the type of the local algebra in the case of two specific spacetime regions, the (infinite) wedges and the (bounded) double cones.

Proposition 10.2 *The local algebra $\mathcal{N}(W)$ belonging to a wedge region W is a type III algebra.*

Proposition 10.3 *If $\{\mathcal{N}(V)\}$ is a dilation invariant net of local von Neumann algebras, then the local algebras $\mathcal{N}(K)$ belonging to the double cones K are type III .*

Let $\mathcal{P}_T$ be the translation subgroup of $\mathcal{P}$. A state ϕ on the quasilocal algebra is called a *translation clustering state* if

$$\lim_{t \to \infty} \phi(x\alpha_{tg}y = \phi(x)\phi(y) \tag{10.11}$$

for all spacelike $g \in \mathcal{P}_T$ and for all local x, y. Note that the vacuum state is a translation clustering state. Furthermore, we have the following result.

Proposition 10.4 *Let $(\mathcal{N}, \{\mathcal{N}(V)\})$ be a net of local von Neumann algebras in the vacuum representation, and assume that the algebras act on the Hilbert space $\mathcal{H}$. If U denotes the unitary representation of the spacetime translation group, and $\mathcal{Z} = \mathcal{N} \cap \mathcal{N}'$ is the center of the quasilocal algebra $\mathcal{N}$, then each element in $\mathcal{Z}$ commutes with $U(g)$ for all translations g.*

This proposition tells us that the translational invariance of the vacuum cannot be spontaneously broken, i.e. the central decomposition of a translation invariant vacuum respects the translation invariance; "the vacuum cannot be a cristal."

Next we wish to list a number of definitions expressing the statistical independence of general C^*-and W^*-algebras. The motivation for such definitions comes from ARQFT: the idea is that local observable algebras should be "independent". More on this idea will be said in Chapter 11. Here we just give the definitions, and we give it for general algebras first. It will be seen that in the specific case of local algebras of ARQFT some definitions turn out to be equivalent.

Let $\mathcal{A}$ and $\mathcal{B}$ be two C^*-subalgebras of the C^*-algebra $\mathcal{C}$ (with common unit). The two algebras $\mathcal{A}$ and $\mathcal{B}$ are said to be *(mutually) commuting* if $XY = YX$ for all $X \in \mathcal{A}$ and $X \in \mathcal{B}$.

Definition 10.3 *Two C^*-subalgebras $\mathcal{A}_1, \mathcal{A}_2$ of the C^*-algebra $\mathcal{C}$ are called C^*-independent if for any state ϕ_1 on $\mathcal{A}_1$ and for any state ϕ_2 on $\mathcal{A}_2$ there is a state ϕ on $\mathcal{C}$ such that*

$$\phi(A) = \phi_1(A) \text{ and } \phi(B) = \phi_2(B) \quad A \in \mathcal{A}_1, B \in \mathcal{A}_2$$

The C^*-independence of the pair $(\mathcal{A}_1, \mathcal{A}_2)$ means that no preparation in any state of the system described by $\mathcal{A}_1$ can exclude any preparation of the system described by $\mathcal{A}_2$. Put it differently: any two states ϕ_1 and ϕ_2 can be prepared by the same preparation process.

One has the following characterization of C^*-independence:

Proposition 10.5 (Schlieder-Roos Theorem) *If the C^*-subalgebras $\mathcal{A}_1$ and $\mathcal{A}_2$ of a C^*-algebra $\mathcal{C}$ are mutually commuting, then the following two conditions are equivalent:*

(i)
$$0 \neq A \in \mathcal{A}_1, 0 \neq B \in \mathcal{A}_2 \quad \text{implies} \quad AB \neq 0 \tag{10.12}$$

(ii) *The pair $(\mathcal{A}_1, \mathcal{A}_2)$ is C^*-independent.*

It has been shown recently that the Schlieder-Roos theorem remains valid without the assumption of mutual commutativity. Specifically we have

Proposition 10.6 *If $\mathcal{A}_1$ and $\mathcal{A}_2$ are two, not necessarily commuting C^*-subalgebras of the C^*-algebra C, then the following two conditions are equivalent*

(i) *$(\mathcal{A}, \mathcal{B})$ are C^*-independent*
(ii) *$\| AB \| = \| A \| \| B \|$ for all $A \in \mathcal{A}_1$ and $B \in \mathcal{A}_2$*

A closely related notion of independence is W^*-independence:

Definition 10.4 *The pair $(\mathcal{M}_1, \mathcal{M}_2)$ of von Neumann algebras is said to be W^*-independent if for any normal state ϕ_1 on $\mathcal{M}_1$ and for any normal state ϕ_2 on $\mathcal{M}_2$ there is a normal state ϕ on $\mathcal{M}$ such that*

$$\phi(A) = \phi_1(A) \text{ and } \phi(B) = \phi_2(B) \quad A \in \mathcal{M}_1, B \in \mathcal{M}_2$$

Note that W^*-independence only makes sense in the category of von Neumann algebras.

Definition 10.5 *The pair $(\mathcal{M}_1, \mathcal{M}_2)$ is said to be W^*-independent in the product sense if for any normal state ϕ_1 on $\mathcal{M}_1$ and for any normal state ϕ_2 on $\mathcal{M}_2$ there is a normal state ϕ on $\mathcal{M}$ that extends both ϕ_1 and ϕ_2 and such that*

$$\phi(AB) = \phi_1(A)\phi_2(B) \quad A \in \mathcal{M}_1, B \in \mathcal{M}_2$$

That is to say, the pair $(\mathcal{M}_1, \mathcal{M}_2)$ is said to be W^*-independent in the product sense if the pair is W^*-independent and the joint extension of the partial normal states is a normal *product state* across the algebras $\mathcal{M}_1$ and $\mathcal{M}_2$. It turns out that W^*-independence is a stronger notion of statistical independence than W^*-independence, see Proposition 10.8 below.

Strict locality is another statistical independence condition formulated for von Neumann algebras.

Definition 10.6 *Let $(\mathcal{M}_1, \mathcal{M}_2)$ be an (ordered) pair of von Neumann subalgebras of the von Neumann algebra $\mathcal{M}$. The pair $(\mathcal{M}_1, \mathcal{M}_2)$ is said to have the independence condition called* strict locality *(or said to be* strictly local*) if for any $A \in \mathcal{P}(\mathcal{M}_1)$ and for any normal state ϕ_2 on $\mathcal{M}_2$ there exists a normal state ϕ on $\mathcal{M}$ such that $\phi(A) = 1$ and $\phi(B) = \phi_2(B)$ for all $B \in \mathcal{M}_2$.*

Strict locality of the pair $(\mathcal{M}_1, \mathcal{M}_2)$ means that no preparation of any state on subsystem $\mathcal{M}_2$ excludes the occurrence of any probability of any proposition $A \in \mathcal{P}(\mathcal{M}_2)$: Let λ be an arbitrary number in the unit interval and A be an arbitrary projection in $\mathcal{P}(\mathcal{M}_1)$. Then for a fixed state ϕ_2 on $\mathcal{M}_2$ there exists states ϕ and ψ on $\mathcal{M}$ such that $\phi(I - A) = 1$ and $\psi(A) = 1$ and which both coincide with ϕ_2 on $\mathcal{M}_2$. The state $\varphi_\lambda = \lambda\psi + (1-\lambda)\phi$ satisfies $\varphi_\lambda(A) = \lambda$. It is not known whether the notion of strict locality is symmetric in $\mathcal{M}_1, \mathcal{M}_2$ in general; it is, however, if $\mathcal{M}_1$ and $\mathcal{M}_2$ commute (see the Proposition 9 in [51]).

Yet another notion of independence is the so-called split property:

Definition 10.7 Two mutually commuting von Neumann subalgebras $\mathcal{M}_1, \mathcal{M}_2$ of the von Neumann algebra $\mathcal{M}$ are said to have the *split property* if there exists a type **I** factor von Neumann algebra $\mathcal{B}(\mathcal{H})$ such that

$$\mathcal{M}_1 \subset \mathcal{B}(\mathcal{H}) \subset \mathcal{M}_2' \qquad (10.13)$$

The relation of the different independence notions is a non-trivial matter with a number of problems still open. However, in the context of ARQFT the situation is somewhat simplified by the following two propositions, which have been recently proven by Florig and Summers [51].

Proposition 10.7 *For a pair of local observable von Neumann algebras $\mathcal{M}(V_1), \mathcal{M}(V_2)$ in a covariant net the following conditions are equivalent:*
(i) *The pair $(\mathcal{M}(V_1), \mathcal{M}(V_2))$ is C^*-independent*
(ii) *The pair $(\mathcal{M}(V_1), \mathcal{M}(V_2))$ is W^*-independent*
(iii) *The pair $(\mathcal{M}(V_1), \mathcal{M}(V_2))$ is strictly local*

Furthermore, we have the following

Proposition 10.8 *The independence condition W^*-independence in the product sense is strictly stronger than W^*-independence.*

It is precisely the violation of Bell's inequality in all normal states for local algebras pertaining to spacelike separated tangent double cones and complementary wedges (Propositions 10.14 and 10.15) that imply the above proposition. This is because a product states satisfy Bell's inequality, hence for the algebras belonging to the specified regions there cannot exist a normal product state. So for the said tangent spacelike separated spacetime regions one is in the remarkable situation, where " ... all normal partial states have normal extensions, *none* of which is allowed to be a product state, and also all partial states have extensions to product states, *none* of which can be normal." ([51] p. 12)

The important condition that is used in the proof of the equivalence stated in Proposition 10.7 is the mutual commutativity of the algebras involved. Little is known about the relation of the different independence conditions in the absence of mutual commutativity. Besides the non-commutative version of the Schlieder-Roos theorem (Proposition 10.6), we mention the following result, recently proven by Florig and Summers in the paper [51].

Proposition 10.9 *Let $\mathcal{M}(V_1)$ and $\mathcal{M}(V_2)$ be not necessarily commuting von Neumann subalgebras of the von Neumann algebra $\mathcal{M}$. If $\mathcal{M}(V_1)$ and $\mathcal{M}(V_2)$ form a strictly local pair, then they are also C^*-independent.*

Also little is known about what conditions (in addition to some independence) on a pair of C^*-or W^*-algebras $(\mathcal{A}, \mathcal{B})$ imply the mutual commutativity of $\mathcal{A}$ and $\mathcal{B}$.

10.2. Bell correlation and Bell's inequality

Definition 10.8 Let $\mathcal{A}$ and $\mathcal{B}$ be two commuting C^*-subalgebras of $\mathcal{C}$ and let ϕ be a state on $\mathcal{C}$. The *Bell correlation* $\beta(\phi, \mathcal{A}, \mathcal{B})$ of the two algebras in the state ϕ is defined by

$$\beta(\phi, \mathcal{A}, \mathcal{B}) \equiv (1/2)\sup \phi(A_1(B_1 + B_2) + A_2(B_1 - B_2)) \qquad (10.14)$$

where the supremum is taken over selfadjoint contractions $A_i \in \mathcal{A}(V_1)$ and $B_i \in \mathcal{A}(V_2)$.

Note that because of mutual commutativity of the algebras the Bell correlation is well-defined, and the products featuring in the correlation are just the ones in terms of which Bell's inequality was formulated in Section 9.4 (see the Proposition 9.12).

Bell's inequality can now be written in the following form:

Definition 10.9

$$\textbf{Bell's inequality:} \qquad \beta(\phi, \mathcal{A}, \mathcal{B}) \le 1 \qquad (10.15)$$

The next proposition gives conditions which are sufficient to imply that Bell's inequality holds.

Proposition 10.10 *Bell's inequality holds in the following cases*
(i) $\mathcal{A}$ *or* $\mathcal{B}$ *is a commutative algebra.*
(ii) ϕ *is a convex combination of product states across* $\mathcal{A}$ *and* $\mathcal{B}$, *i.e. there are states* ϕ_1^i *and* ϕ_2^i *over* $\mathcal{A}$ *and* $\mathcal{B}$ *respectively such that* $\phi = \sum_i \lambda_i \phi_1^i \phi_2^i$ *for some constants* λ_i *such that* $\sum_i \lambda_i = 1$.
(iii) *The restriction of* ϕ *to either* $\mathcal{A}$ *or* $\mathcal{B}$ *is a pure state.*

Proof: (i) Because of the symmetrical role of the algebras we may assume that $\mathcal{A}$ is commutative. In this case all the four elements

$$A_{+,+} = \frac{1}{4}(I + A_1)(I + A_2)$$

$$A_{-,-} = \frac{1}{4}(I - A_1)(I - A_2)$$

$$A_{+,-} = \frac{1}{4}(I + A_1)(I - A_2)$$

$$A_{-,+} = \frac{1}{4}(I - A_1)(I + A_2)$$

are positive, and one can compute directly that

$$\begin{aligned}\beta(\phi, \mathcal{A}(V_1), \mathcal{A}(V_2)) &= \phi(A_{+,+}B_1 + \phi(A_{+,-}B_2) \\ &\quad - \phi(A_{-,+}B_2) - \phi(A_{-,-}B_1) \\ &\leq \phi((A_{+,+} + A_{+,-} + A_{-,+} + A_{-,-}, I) = \phi(I) = 1\end{aligned}$$

(ii) It is enough to see that if ϕ itself is a product state with some states φ and ω, then the statement is true. One can write

$$\phi(A_1(B_1 + B_2) + A_2(B_1 - B_2)) = \\ \omega(A_1)\varphi(B_1) + \omega(A_1)\varphi(B_2) + \omega(A_2)\varphi(B_1) - \omega(A_2)\varphi(B_2)$$

with contractions $A_i \in \mathcal{A}$ and $B_i \in \mathcal{B}$. Since

$$|\omega(A_i)| \leq 1$$
$$|\varphi(B_i)| \leq 1$$

the statement is a consequence of the following simple lemma.

Lemma: If a_1, a_2 and b_1, b_2 are real numbers in the interval $[0, 1]$, then we have the inequality

$$|\frac{1}{2}(a_1 b_1 + a_1 b_2 + a_2 b_1 - a_2 b_2)| \leq 1$$

Proof of Lemma:

$$\begin{aligned}&|\frac{1}{2}(a_1 b_1 + a_1 b_2 + a_2 b_1 - a_2 b_2)| \\ &\leq \frac{1}{2}|a_1||b_1 + b_2| + \frac{1}{2}|a_2||b_1 - b_2| \\ &\leq \frac{1}{2}|b_1 + b_2| + \frac{1}{2}|b_1 - b_2|\end{aligned}$$

One can now check explicitly that the right hand side of the last inequality is $b_1, -b_1, b_2$ or $-b_2$ depending on the signs of $(b_1 + b_2)$ and $(b_1 - b_2)$.

(iii) The purity of an arbitrary state ϕ on a C^*-algebra $\mathcal{A}$ is equivalent to the condition that if the functional ω on $\mathcal{A}$ is such that $-\phi \leq \omega \leq \phi$, then ω is of the form $\lambda\phi$ for some $-1 \leq \lambda \leq 1$. Since the functional

$$\omega_B(A) \equiv \phi(AB) \quad A \in \mathcal{A}, B \in \mathcal{B}$$

is such that
$$-\phi(A) \leq \omega_B(A) \leq \phi(A) \quad A \in \mathcal{A}$$
the purity of ϕ on $\mathcal{A}$ implies that for every $B \in \mathcal{B}$ there exists a number $-1 \leq \varphi(B) \leq 1$ such that
$$\phi(AB) = \omega_B(A) = \varphi(B)\phi(A) \quad A \in \mathcal{A}, B \in \mathcal{B}$$
That is ϕ is a product state, so the statement follows from (ii). □

Standard derivations of Bell's inequality in a hidden variable framework make two assumptions: that expectation values of the quantum observables can be represented by a classical probability theory in the sense that the quantum expectation values can be obtained as expectation values computed in a classical (i.e. commutative) probability calculus, *and* that the classical representation of the quantum observables/expectation values is "local", which is commonly interpreted as the requirement that the hidden state be a product state in the sense formulated in condition (ii) of the above proposition. The proposition above shows that *any one* of these two conditions is already sufficient to entail Bell's inequality; that is to say, once one has assumed that the quantum observables are represented by random variables on a classical probability space, one is committed to Bell's inequality, any locality condition is redundant. Hence the statement (i) in Proposition 10.10 can be interpreted by saying that Bell's inequality puts an upper bound on the strength of correlations that can be modelled by classical (i.e. commutative) probability spaces. Thus, if in an experimental situation one would find correlations not satisfying Bell's inequality, then one would have to conclude that the correlations in question cannot be modelled by classical probability spaces. The proposition below puts an upper bound on the strength of correlations that can be modelled by possibly non-commutative C^*-algebras:

Proposition 10.11 *For any state ϕ and for arbitrary C^*-algebras and $\mathcal{A}$ and $\mathcal{B}$ we have*

$$\beta(\phi, \mathcal{A}, \mathcal{B}) \leq \sqrt{2} \qquad (10.16)$$

Proof: Let $(\mathcal{H}_\phi, \pi_\phi, \Omega_\phi)$ be the GNS triplet determined by the state ϕ (see Section 6.1). Then for each $B \in \mathcal{B}$ such that $I \leq B \leq I$ the formula

$$\phi_B(X) \equiv \phi(XB) \qquad X \in \mathcal{A} \qquad (10.17)$$

defines a linear functional ϕ_B on $\mathcal{A}$ such that $-\phi \leq \phi_B \leq \phi$. Therefore there exists a unique $B' \in \pi_\phi(\mathcal{A})'$ such that

$$\phi_B(X) = \langle \Omega_\phi, \pi_\phi(X) B' \Omega_\phi \rangle \qquad X \in \mathcal{A} \qquad (10.18)$$

Let B_1', B_2' be the elements in the commutant $\pi_\phi(\mathcal{A})'$ obtained in this way by inserting B_1 and B_2 into the formula (10.17), and let us deefine

$$A \equiv \frac{1}{2}\pi_\phi(A_1 + iA_2)$$

$$B \equiv \frac{1}{2\sqrt{2}}(B_1' + B_2' + i(B_1' + B_2'))$$

Then we have

$$A^*A + AA^* = \frac{1}{2}\pi_\phi(A_1^2 + A_2^2) \leq 1$$

$$B^*B + BB^* = \frac{1}{2}(B_1'^2 + B_2'^2) \leq 1$$

Furthermore

$$\begin{aligned}
\sqrt{2}\beta(\phi, \mathcal{A}, \mathcal{B}) &= 4\mathrm{Re}\langle \Omega_\phi, A^*B\Omega_\phi\rangle \\
&= 2\mathrm{Re}\langle A\Omega_\phi, B\Omega_\phi\rangle + 2\mathrm{Re}\langle B^*\Omega_\phi A^*\Omega_\phi\rangle \\
&= \|A\Omega_\phi\|^2 + \|B\Omega_\phi\|^2 - \|(A-B)\Omega_\phi\|^2 \\
&\quad + \|B^*\Omega_\phi\|^2 + \|A^*\Omega_\phi\|^2 - \|(B^* - A^*)\Omega_\phi\|^2 \\
&\leq \langle \Omega_\phi, A^*A + AA^* + B^*B + BB^*)\Omega_\phi\rangle \leq \sqrt{2}
\end{aligned}$$

□

The inequality (10.16) in the above proposition may be called "noncommutative Bell's inequality", since, if in some experiment one would find that $\beta(\phi, \mathcal{A}, \mathcal{B}) > \sqrt{2}$, this would mean that the C^*-algebraic framework is unsuitable to describe the correlations obtained in the given experiment. A violation of $\beta(\phi, \mathcal{A}, \mathcal{B}) \leq \sqrt{2}$ is *not* apriori impossible but so far no experimental results showing the limits of C^*-algebraic framework are known, and consequently, the C^*-algebraic framework is the one in which the subsequent discussions will be carried out. Thus we make the following stipulation.

Definition 10.10 Two commuting C^*-subalgebras $\mathcal{A}, \mathcal{B}$ of the C^*-algebra $\mathcal{C}$ are said to *maximally violate Bell's inequality in state ϕ on $\mathcal{C}$* if

$$\beta(\phi, \mathcal{A}, \mathcal{B}) = \sqrt{2}$$

The fact that through (1) in Proposition 10.10 and through Proposition 10.9 above it is possible to exclude certain mathematical probability theories as models of experimentally obtainable correlations is significant, both philosophically and physically. It is *not* evident that one can have a relatively simple means (an inequality) to falsify certain mathematical theories of probability as an appropriate model of statistical events, and it is not evident, further, that if one has such a test in principle, then the test, the particular falsifier, is empirically as accessible as the Bell correlation is. If the Bell correlation were a hopelessly complicated expression one would probably not have attempted checking its value in experimnets.

The Bell correlation has been tested on quantum systems empirically, and all the evidence suggest that the correlation violates Bell'sinequality but remains within the bound $\sqrt{2}$ (see the bibliographic remarks).

10.3. Violation of Bell's inequality in quantum field theory

Proposition 10.10 shows that there exist an (infinite) number of states for which Bell's inequality holds. If, however, neither $\mathcal{A}$ nor $\mathcal{B}$ is commutative, then, under some additional hypotheses that are met in ARQFT, Bell's inequality cannot be satisfied in all states; in fact there exist states in which the Bell correlation takes on its *maximal* value ($\sqrt{2}$), as the next proposition shows.

Proposition 10.12 *Let $(\mathcal{M}, \mathcal{N})$ be a pair of commuting von Neumann algebras acting on the Hilbert space $\mathcal{H}$ such that if $A \in \mathcal{M}, B \in \mathcal{N}$ and $AB = 0$ implies either $A = 0$ or $B = 0$ ("Schlieder property"). Then if neither $\mathcal{M}$ nor $\mathcal{N}$ is commutative, then there exists a normal state ϕ such that $\beta(\phi, \mathcal{M}, \mathcal{N}) = \sqrt{2}$.*

Proof: In the proof we shall use the following

Lemma: If $\mathcal{M}$ is a non-commutative von Neumann algebra, then there exist two non-commuting projections P and Q such that $\| [P,Q] \| = \frac{1}{2}$.

Proof of Lemma: Assume that $\mathcal{M}$ acts on the Hilbert space $\mathcal{H}$, and let us assume that there exists a partial isometry V in $\mathcal{M}$ such that the two projections $A = VV^*$ and $B = V^*V$ are orthogonal. Then $\mathcal{H}$ can be written as

$$\mathcal{H}_A \oplus \mathcal{H}_B \oplus \mathcal{K}$$

where $\mathcal{K}$ is the orthogonal complement of $A + B$. One can define two operators P and Q in $\mathcal{M}$ by

$$P(\xi_A + \xi_B + \xi_\mathcal{K}) =$$
$$\frac{1}{\sqrt{2}}(vv^*\xi_A + v\xi_B) + \frac{1}{\sqrt{2}}(v^*\xi_A + vv^*\xi_B)$$
$$Q(\xi_A + \xi_B + \xi_\mathcal{K}) =$$
$$\frac{1}{\sqrt{2}}(vv^*\xi_A - iv\xi_B) + \frac{1}{\sqrt{2}}(iv^*\xi_A + vv^*\xi_B)$$

where the vectors ξ_A, ξ_B and ξ_K are from the subspaces indicated. P and Q can then be written compactly as

$$P \begin{pmatrix} \xi_A \\ \xi_B \end{pmatrix} = \frac{1}{\sqrt{2}} \begin{pmatrix} vv^* & v \\ v^* & v^*v \end{pmatrix} \begin{pmatrix} \xi_A \\ \xi_B \end{pmatrix}$$

and

$$Q \begin{pmatrix} \xi_A \\ \xi_B \end{pmatrix} = \frac{1}{\sqrt{2}} \begin{pmatrix} vv^* & -iv \\ iv^* & v^*v \end{pmatrix} \begin{pmatrix} \xi_A \\ \xi_B \end{pmatrix}$$

Since the matrices

$$\frac{1}{\sqrt{2}} \begin{pmatrix} vv^* & v \\ v^* & v^*v \end{pmatrix}$$

and

$$\frac{1}{\sqrt{2}} \begin{pmatrix} vv^* & -iv \\ iv^* & v^*v \end{pmatrix}$$

of P and Q are selfadjoint and are easily seen to be idempotent, P and Q are projections; furthermore, one can check by an explicit calculation that

$$(PQ - QP)\xi_B = \frac{1}{2} iv\xi_B$$

Since v is a partial isometry and ξ_B is in the initial space of v it follows that

$$\| (PQ - QP)\xi_B \| = \frac{1}{2} \| iv\xi_B \|$$

which implies $\| (PQ - QP) \| = \frac{1}{2}$. Thus P and Q are the two projections whose existence was to be shown. To prove the Lemma, one has then to show that if $\mathcal{M}$ is non-commutative, then it contains a partial isometry v such that vv^* and v^*v are orthogonal projections. If $\mathcal{M}$ is non-commutative, then it contains two non-commuting projections R and S, say. The Hilbert space $\mathcal{H}$ can then be written as

$$\mathcal{H} = \mathcal{H}_R \oplus \mathcal{H}_{R^\perp}$$

and R and S can be written as

$$R = \begin{pmatrix} R & 0 \\ 0 & 0 \end{pmatrix} \qquad S = \begin{pmatrix} S_{11} & T \\ T^* & S_{22} \end{pmatrix}$$

with some operators S_{11}, S_{22} and T. If R and S do not commute, then $T \neq 0$, and T then has a polar decomposition $T = V(T^*T)^{1/2}$, which can be written on the space $\mathcal{H}_R \oplus \mathcal{H}_{R^\perp}$ as

$$T = \begin{pmatrix} 0 & T \\ 0 & 0 \end{pmatrix} = \begin{pmatrix} 0 & v \\ 0 & 0 \end{pmatrix} \begin{pmatrix} 0 & 0 \\ 0 & (T^*T)^{1/2} \end{pmatrix}$$

with
$$V = \begin{pmatrix} 0 & v \\ 0 & 0 \end{pmatrix}$$

where v is a partial isometry. The partial isometry V has then the property that
$$VV^* = \begin{pmatrix} vv^* & 0 \\ 0 & 0 \end{pmatrix}$$

and
$$V^*V = \begin{pmatrix} 0 & 0 \\ 0 & v^*v \end{pmatrix}$$

are orthogonal projections.

If P is an arbitrary projection on $\mathcal{H}$, then $2P-I$ is a selfadjoint contraction. Let $P_i \in \mathcal{M}$, $Q_j \in \mathcal{N}$ ($i,j = 1,2$) be projections. Then $A_i \equiv 2P_i - I$ and $B_j \equiv 2Q_j - I$ ($i,j,= 1,2$) are selfadjoint contractions. Introducing the notation

$$Z \equiv A_1(B_1 + B_2) + A_2(B_1 - B_2) \tag{10.19}$$

one can calculate explicitly

$$Z^2 = 4 + 16([P_1, P_2][Q_1, Q_2]) \tag{10.20}$$

There exists a normal state φ such that

$$|\varphi([P_1, P_2][Q_1, Q_2])| = \|[P_1, P_2][Q_1, Q_2]\| \tag{10.21}$$

and one may assume that

$$\varphi([P_1, P_2][Q_1, Q_2]) \geq 0 \tag{10.22}$$

for if this were not the case, then interchanging A_1, A_2 we could change the sign of the left hand side of eq. (10.22). So there exists a state φ such that

$$\varphi(Z^2) = 4 + 16 \|[P_1, P_2][Q_1, Q_2]\| \tag{10.23}$$

and it follows that

$$\|A_1(B_1 + B_2) + A_2(B_1 - B_2)\| = 2\sqrt{1 + 4\|[P_1, P_2][Q_1, Q_2]\|} \tag{10.24}$$

There exists a normal state ϕ on $\mathcal{B}(\mathcal{H})$ such that

$$\frac{1}{2} |\phi(A_1(B_1 + B_2) + A_2(B_1 - B_2))| = \sqrt{1 + 4\|[P_1, P_2][Q_1, Q_2]\|} \tag{10.25}$$

By the Schlieder-Roos theorem (Proposition 10.5 and Proposition 10.6) the Schlieder property implies

$$\|[P_1, P_2][Q_1, Q_2]\| = \|[P_1, P_2]\| \|[Q_1, Q_2]\| \tag{10.26}$$

and by the Lemma one can find two pairs of projections P_1, P_2 and Q_1, Q_2 such that

$$\| [P_1, P_2] \| = \frac{1}{2}$$
$$\| [Q_1, Q_2] \| = \frac{1}{2}$$

so the statement in the proposition follows. □

Let $(\mathcal{A}, \mathcal{A}(V), V \subset M)$ be a net of local algebras in the sense of ARQFT described in the previous section. Since the local algebras $\mathcal{A}(V_1)$ and $\mathcal{A}(V_2)$ pertaining to spacelike separated spacetime regions V_1 and V_2 mutually commute by the axiom of microcausality and they also have a common unit, the Bell correlation $\beta(\phi, \mathcal{A}(V_1), \mathcal{A}(V_2))$ between the local algebras $\mathcal{A}(V_1)$ and $\mathcal{A}(V_2)$ exists as a meaningful quantity:

$$\beta(\phi, \mathcal{A}(V_1), \mathcal{A}(V_2)) \equiv \frac{1}{2}\sup\phi(A_1(B_1 + B_2) + A_2(B_1 - B_2)) \qquad (10.27)$$

(ϕ being an arbitrary state on the quasilocal algebra $\mathcal{A}$, and the supremum is taken over selfadjoint contractions $A_i \in \mathcal{A}(V_1)$ and $B_i \in \mathcal{A}(V_2)$.)

Let $(\mathcal{N}, \{\mathcal{N}(V)\}, V \subset M)$ be a covariant net of local von Neumann algebras in ARQFT described in the previous section. Since the local algebras in ARQFT are non-commutative and the local algebras belonging to typical spacelike separated spacetime regions (such as strictly spacelike separated regions or spacelike separated tangent double cones and wedges) are also C^*-independent – and thus have the Schlieder property by the Schlieder-Roos theorem – we have as a consequence of Proposition 10.12 the following:

Proposition 10.13 *Let V_1, V_2 be two spacelike separated spacetime regions such that the local von Neumann algebras $\mathcal{N}(V_1)$ and $\mathcal{N}(V_2)$ are C^*-independent. Then there is a normal state ϕ such that $\beta(\phi, \mathcal{A}(V_1), \mathcal{A}(V_2)) = \sqrt{2}$.*

This proposition tells us that Bell's inequality is violated in some normal state for some observables that belong to certain typical spacelike separated spacetime regions, however "far apart" these regions might be.

But much more is in fact true. Summers and Werner have shown that the violation of Bell's inequality in ARQFT is (i) typical, (ii) maximal and (iii) generic.

That the violation is typical means that Bell's inequality is violated in *every* normal state for local algebras pertaining to characteristic *tangent* spacetime regions. These regions are: *all* spacelike separated *tangent* double cones, *all* spacelike separated *tangent* wedges. Maximality of the typical violation means that the Bell correlation attains its maximal value ($\sqrt{2}$) in the normal states for the mentioned spacetime regions. By the genericity of the violation is meant that the maximal and typical violation is not a peculiar feature of a single, particular net of local algebras, but it is characteristic of a number of field theories that, in addition to the minimal set of axioms of isotony, locality and Poincare covariance, satisfy some further conditions. The propositions and theorems that altogether embody the typical and maximal violation of Bell's inequality vary in their generality and applicability, and they were discovered step-by-step in a series of papers in the past decade. Calling a local field theory "good" if, in addition to the minimal set of postulates (i)-(v), it has the property of weak additivity and is such that both wedge duality and the cone cyclicity of the vacuum hold, one may say that "Bell's inequalities are maximally and typically violated in every good ARQFT". To be more precise, one has the following two typical results.

Proposition 10.14 *Let $\{\mathcal{N}(V)\}$ be a local net in an irreducible vacuum representation and assume that the net satisfies wedge duality, and cone cyclicity (which is the case if $\{\mathcal{N}(V)\}$ is associated with a Wightman field). Then Bell's inequality is maximally violated in all normal states for all algebras pertaining to complementary wedge regions W and W'.*

Proposition 10.15 *Let $\{\mathcal{N}(V)\}$ be a dilatation-invariant local net in an irreducible vacuum representation such that weak additivity and wedge duality hold for the net (which is the case if the net $\{\mathcal{N}(V)\}$ is associated with a Wightman field). Then Bell's inequality is maximally violated in all normal states for all algebras pertaining to spacelike separated tangent double cones.*

10.4. Superluminal correlations in quantum field theory

Let $(\mathcal{N}, \{\mathcal{N}(V)\})$ be a covariant net of (strictly) local von Neumann algebras $\mathcal{N}(V)$ in the sense of algebraic quantum field theory described in Section 10.1, and let V_1 and V_2 be two spacelike separated spacetime regions. If $A \in \mathcal{N}(V_1)$ and $B \in \mathcal{N}(V_2)$ are two projections in the respective algebras and ϕ is a state on the quasilocal algebra $\mathcal{N}$, then it can happen very well that

$$\phi(AB) > \phi(A)\phi(B) \qquad (10.28)$$

If (10.28) is the case, then we say that there is *superluminal correlation* between A and B in state ϕ.

A typical example of superluminal correlation is the one predicted by the vacuum state ϕ_0: If V_1 and V_2 are two spacelike separated tangent double cone regions, or two spacelike separated complementary wedge regions in the Minkowski spacetime, then

$$\phi_0(AB) > \phi_0(A)\phi_0(B) \qquad (10.29)$$

for some projections $A \in \mathcal{A}(V_1), B \in \mathcal{A}(V_2)$.

The existence of such A, B is a consequence of the fact that the vacuum state violates Bell's inequality for the said regions in "every" field theory. This is because a product state satisfies Bell's inequality (Proposition 10.10), hence ϕ_0 cannot be a product state across the algebras $\mathcal{N}(V_1), \mathcal{N}(V_2)$, and it follows that there exist selfadjoint contractions $X \in \mathcal{N}(V_1), Y \in \mathcal{N}(V_2)$ such that

$$\phi_0(XY) \neq \phi_0(X)\phi_0(Y)$$

which implies that

$$\phi_0(P_1 P_2) \neq \phi_0(P_1)\phi_0(P_2)$$

for some spectral projections P_1, P_2 of X and Y respectively, hence either

$$\phi_0(P_1 P_2) > \phi_0(P_1)\phi_0(P_2)$$

or

$$\phi_0(P_1^\perp P_2) > \phi_0(P_1^\perp)\phi_0(P_2)$$

holds.

Unless one takes the position that correlations need not be explained at all, a position taken by Van Fraassen for example [161], if one sees a probabilistic correlation, one would like to say either of the following

1. There exists a direct causal connection between the correlated events.
2. There exists a probabilistic common cause of the correlation.

In fact, in the case of *superluminal* correlations such as the one predicted by ARQFT, one would *not* like to say 1. – and consider it true, too, since spacelike separated events are not supposed to causally influence each other. Yet, option 1. is not a priori impossible, for it can happen that ARQFT does not comply with the no-action-at-a-distance principle, despite the fact that this theory was constructed precisely with the aim of creating a quantum theory that complies with the no-action-at-a-distance principle: There is no apriori assurance that the axioms a net of local observable algebras is required to satisfy do indeed exclude unwanted causal connections. However, to claim that there is (or that there is no) causal connection between spacelike separated events, one has to specify "causal connection" or "causal independence" in terms of ARQFT precisely enough to be able to prove absence/presence of a causal link. So the presence of superluminal correlations in ARQFT leads naturally to the problem of independence of the correlated algebras. We have seen in Section 10.1 that the spacelike separated algebras satisfy a number of independence conditions. The next chapter adds two more to these: logical independence and counterfactual probabilistic independnce. It turns out that the spacelike separated local algebras in ARQFT satisfy these independence conditions as well.

These "negative" results on option 1 as explanation of the superluminal correlation leave one with option 2. With the problem namely, if there is a common cause explanation of the correlation. This problem is taken up in Chapter 12. We shall see that this problem is entirely open.

10.5. Bibliographic notes

The axioms of algebraic quantum field theory were first systematically formulated in 1964 in [57]. The theory has since reached maturity and is summarized in the monographs of Haag [56] and Horuzhy [72]. For the relation of the Wightman axioms to algebraic quantum field theory see [48]. The two propositions on the type of local algebras can be found in [47]. The notion of C^*-independence of local observable algebras was introduced by Haag and Kastler in [57] in the framework of algebraic quantum field theory. The different notions of independence, including statistical independence are reviewed in [151]. The notion of Bell correlation was introduced by Summers and Werner in [145], its bounds and value in quantum field theory, thus violation of Bell's inequality in field theory, have been extensively investigated by Landau [91], and by Summers and Werner in a number of papers [145], [146], [147], [148] [149], [152] (see [151] and [150] for reviews). The value of the Bell correlation was checked experimentally. The idea of the experiment is outlined in the review [40], the results of the experiment were published in [6], [7] and [8]. The commonly accepted interpretation of the experimental results is that Bell's inequality is violated by Nature. For a dissenting view see [156].

CHAPTER 11

Independence in quantum logic approach

An important notion in physics is the concept of *independence* of physical systems. One typically encounters the problem of independence of systems in the situation where S is a physical system and S_1, S_2 are two subsystems of S. The problem of independence comes then either in the form of the need to decide whether S_1 and S_2 are independent, or in the form of the need to impose an independence condition on S_1, S_2, as part of creating a suitable model of the systems involved. Relativistic quantum field theory is a case in point. We have seen in the last chapter that, on the one hand, one imposes the local commutativity (microcausality) condition on the net of local algebras, which, together with the other axioms imply other (statistical) independence conditions; on the other hand, the existence of probabilistic correlations between distant (spacelike separated) projections raises the suspicion that the (spacelike separated) local algebras are not, after all "independent" in some sense in which one expects them to be. Obviously, it is then of interest to clarify the independence relations between two subsystems of a larger quantum system, and to do this one needs intuitively and physically justifiable, and mathematically operational concepts of independence.

Now the notion of independence is not a theory-independent one. Rather, it is to be expected that in different models of even one and the same physical system different independence conditions are natural and suitable. As we have seen in the previous chapter, if the quantum systems are represented by C^*-or W^*-algebras, then one can formulate different, nonequivalent notions of statistical independence notions, such as C^*-independence, W^*-independence, strict locality, W^*-independence in the product sense and split property. In the quantum logic approach to quantum mechanics, a quantum system S and two of its subsystems S_1, S_2 are modeled by the orthomodular lattice of projections $\mathcal{P}(\mathcal{M})$ of the von Neumann algebra $\mathcal{M}$ and the projection lattices $\mathcal{P}(\mathcal{M}_1), \mathcal{P}(\mathcal{M}_2)$ of two von Neumann subalgebras $\mathcal{M}_1, \mathcal{M}_2$ of $\mathcal{M}$. These lattices represent the logical structure of (some of) the empirically testable propositions regarding the observable quantities of the systems, if the observables are represented by (the selfadjoint part of) the algebras $\mathcal{M}_1, \mathcal{M}_2$ and $\mathcal{M}$, and if the states of the system are given by the state spaces of $\mathcal{M}_1, \mathcal{M}_2$ and

$\mathcal{M}$. It is natural then to ask for a notion of *logical independence* of the two von Neumann sublattices $\mathcal{P}(\mathcal{M}_1), \mathcal{P}(\mathcal{M}_2)$. The aim of this chapter is to formulate definitions of independence for a pair of lattices $\mathcal{P}(\mathcal{M}_1), \mathcal{P}(\mathcal{M}_2)$ in lattice theoretic terms and to characterize the pairs of independent lattices.

We shall consider two kinds of independence in detail: *logical (semantic)* and *counterfactual probabilistic* independence, and both independence notions will be related to notions of statistical independence.

In Section 11.1 we first motivate the main definition of the section, the Definition 11.6, which expresses that no (non-trivial) proposition in a von Neumann lattice $\mathcal{P}(\mathcal{M}_1)$ implies or is implied by any proposition in the von Neumann lattice $\mathcal{P}(\mathcal{M}_2)$, $\mathcal{M}_1, \mathcal{M}_2$ being von Neumann subalgebras of the von Neumann algebra $\mathcal{M}$. A result of Murray and von Neumann is then recalled that implies that $\mathcal{P}(\mathcal{M}_1), \mathcal{P}(\mathcal{M}_2)$ are logically independent if $\mathcal{M}_1$ is a factor von Neumann algebra and $\mathcal{P}(\mathcal{M}_1), \mathcal{P}(\mathcal{M}_2)$ commute (Proposition 11.3). A proof of this is given in the special case where $\mathcal{M}_1$ is a subfactor in a finite factor $\mathcal{M}$ and $(\mathcal{M}_1, \mathcal{M}_2)$ is a commuting pair of subalgebras (Proposition 11.4 and its proof). Also, logical independence will be related in this section to three statistical independence conditions called C^*-independence, W^*-independence and strict locality. It turns out that logical independence of $\mathcal{P}(\mathcal{M}_1), \mathcal{P}(\mathcal{M}_2)$ is equivalent to the C^*-independence of $(\mathcal{M}_1, \mathcal{M}_2)$ *if* the algebras $\mathcal{M}_1, \mathcal{M}_2$ mutually commute (Proposition 11.5). Since it is known that W^*-independence implies strict locality and strict locality implies C^*-independence *if* $(\mathcal{M}_1, \mathcal{M}_2)$ is a commuting pair, it follows that for commuting algebras $\mathcal{M}_1, \mathcal{M}_2$ the lattices $\mathcal{P}(\mathcal{M}_1), \mathcal{P}(\mathcal{M}_2)$ are logically independent if the pair $(\mathcal{M}_1, \mathcal{M}_2)$ is either W^*-independent or fulfils the independence condition strict locality (Proposition 11.7). No relation between strict locality and C^*-independence is known if $\mathcal{M}_1, \mathcal{M}_2$ are not commuting general von Neumann algebras, however. But it will be shown that, whether or not $\mathcal{M}_1$ and $\mathcal{M}_2$ commute, the C^*-independence of $\mathcal{M}_1, \mathcal{M}_2$ implies logical independence of $\mathcal{P}(\mathcal{M}_1), \mathcal{P}(\mathcal{M}_2)$ if $\mathcal{M}$ is a *finite dimensional* full matrix algebra (Proposition 11.6), and that $\mathcal{P}(\mathcal{M}_1), \mathcal{P}(\mathcal{M}_2)$ are logically independent if $(\mathcal{M}_1, \mathcal{M}_2)$ is a strictly local or W^*-independent pair (Proposition 11.8). It would be desirable to have suitable further general characterizations of logically independent sub-quantum lattices, but this problem remains largely open. Particular open questions related to the characterization problem are formulated explicitly in Problems 11.1, 11.2, 11.3, 11.4, 11.5 and 11.6.

In Section 11.2 a counterfactual analysis of the problem of probabilistic causal independence between spacelike separated events in algebraic relativistic quantum field theory (ARQFT) is proposed. The analysis is based on Lewis' idea of chancy causation and on Stalnaker's possible world semantics of counterfactuals. Lewis' theory is specified in terms of ARQFT and a precise definition of counterfactual probabilistic causal dependence between events understood as projections in the von Neumann algebras localized in spacelike separated spacetime regions is given (Definition 11.10). The general problem of whether there are spacelike separated local observable algebras containing projections with conterfactual probabilistic dependence between them remains open (Problem 11.6)). It will be shown, however, that if the two von Neumann algebras representing the local observables belonging to spacelike separated spacetime regions have the independence property of C^*-independence, then they contain no projections with counterfactual probabilistic causal dependence between them (Proposition 11.9). Since Bell's inequality *is* violated in quantum field theory even for local algebras having the independence property of C^*-independence, it follows then that violation of Bell's inequality in ARQFT does *not* imply presence of superluminal causation in the counterfactual sense in ARQFT (Proposition 11.10).

11.1. Logical independence in quantum logic

11.1.1. LOGICAL NOTIONS OF INDEPENDENCE

As it was seen in Chapter 5, in the semantic approach to the physical theory the physical theory is represented by a semi interpreted language $(\mathcal{L}, \mathcal{F}, h, \Gamma)$, and the semantic notions, like the truth of a sentence $\alpha \in \mathcal{F}$, the semantic entailment $\alpha \models \beta$ etc. are defined with the help of the map h that assigns to every elementary sentence those states in Γ that make true the sentence. It is rather natural to assume that if S is represented by the language $(\mathcal{L}, \mathcal{F}, h, \Gamma)$, then the two subsystems S_1 and S_2 are described by two "sub-languages" $(\mathcal{L}, \mathcal{F}_1, h, \Gamma_1)$ and $(\mathcal{L}, \mathcal{F}_2, h, \Gamma_2)$, with $\mathcal{F}_1, \mathcal{F}_2 \subseteq \mathcal{F}$ being two closed (with respect to the logical operations in $\mathcal{F}$) subsets of sentences and Γ_1, Γ_2 being the state space of S_1 and S_2.

In this semantic framework the independence of S_1, S_2 should be formulated as the "logical independence" of the two sets $\mathcal{F}_1, \mathcal{F}_2$ of the empirically checkable, meaningful statements on the respective systems. The natural notion of independence that comes to mind is that "no (non-trivial) proposition in $\mathcal{F}_1$ should imply any (non-trivial) proposition in $\mathcal{F}_2$ and conversely, no (non-trivial) proposition in $\mathcal{F}_2$ should imply any (non-trivial) proposition in $\mathcal{F}_1$".

There are several options to implement this idea of independence. The first is to take "implies" in the sense of semantic entailment:

Definition 11.1 $\mathcal{F}_1, \mathcal{F}_2$ are *semantically independent* if

$$\text{semantic independence:} \quad \alpha \not\models \beta \quad \beta \not\models \alpha \quad (11.1)$$

for any non-trivial statements $\alpha \in \mathcal{F}_1, \beta \in \mathcal{F}_2$ where $\models$ is a relation of semantic entailment between the sentences.

Assume now that there exists an implication connective $\Rightarrow$ in the set of sentences $\mathcal{F}$, a two-place connective that expresses formally certain features of the implication "if α then β". Then a second implementaiton of the independence idea can be what may be called "independence with respect to an implication connective $\Rightarrow$", or "$\Rightarrow$-independence" for short. The idea of $\Rightarrow$-independence of $\mathcal{F}_1, \mathcal{F}_2$ is that no (non-trivial) proposition in $\mathcal{F}_1$ "implies" any (non-trivial) proposition in $\mathcal{F}_2$, and, conversely, no proposition in $\mathcal{F}_2$ "implies" any proposition in $\mathcal{F}_1$, where "does not imply" is now taken in the sense that the inference (with respect to $\Rightarrow$) between the elements of $\mathcal{F}_1$ and $\mathcal{F}_2$ is not a tautology.

Definition 11.2 $\mathcal{F}_1, \mathcal{F}_2$ are called $\Rightarrow$-*independent* if

$$\Rightarrow \text{-independence:} \quad (\alpha \Rightarrow \beta) \text{ and } (\beta \Rightarrow \alpha) \text{ are not tautologies} \quad (11.2)$$

for any (non-trivial) $\alpha \in \mathcal{F}_1, \beta \in \mathcal{F}_2$

Finally, another implementation of the independence can be that no statement $\beta \in \mathcal{F}_2$ follows from the statements in $\mathcal{F}_1$ in the sense that both $\mathcal{F}_1 \cup \{\sim \alpha\}$ and $\mathcal{F}_1 \cup \{\alpha\}$ are satisfiable sets of statements, and conversely, no statement $\alpha \in \mathcal{F}_1$ follows from the statements in $\mathcal{F}_2$ in this sense.

Definition 11.3 For any set $\mathcal{F}_0 \subseteq \mathcal{F}$ of statements let $\sim \mathcal{F}_0$ denote the set of statements

$$\mathcal{F}_0 \equiv \{\sim \gamma \mid \gamma \in \mathcal{F}_0\}$$

$\mathcal{F}_1, \mathcal{F}_2$ are called *logically independent* if

$$\mathcal{F}_1 \cup \mathcal{F}_2, \quad \sim \mathcal{F}_1 \cup \mathcal{F}_2, \quad \mathcal{F}_1 \cup \sim \mathcal{F}_2, \quad \sim \mathcal{F}_1 \cup \sim \mathcal{F}_2$$

are all satisfiable sets of sentences.

Having these independence definitions, it is natural to ask whether they are different. Clearly, as long as the key elements in the definitions ($\models$ in (11.1), $\Rightarrow$ in (11.2) and the concept of satisfaction in Definition (11.3)) are not linked to each other formally, the three independence notions remain unrelated. So the first task is to establish a connection between these logical concepts. This can be achieved by considering them in terms of the propositional system that the semi-interpreted language determines.

Recall that in the case of quantum mechanics the set of equivalence classes $\mathcal{F}_\leftrightarrow^q$ of the statements $\mathcal{F}^q$ (with respect to the relation "α is true if and only if β is true') was shown in Chapter 5 to be isomorphic with the orthomodular lattice $\mathcal{P}(\mathcal{H})$ of the set of all projections on the Hilbert space $\mathcal{H}$ (more generally, with the orthomodular lattice $\mathcal{P}(\mathcal{M})$ of projections of a von Neumann algebra $\mathcal{M}$, see the end of Section 6.2). The equivalence was established by the map $|h^q|$ that assigned every equivalence class those states that make true the statements in the equivalence class. Under this identification a sentence in $\mathcal{F}$ is satisfiable if it is represented by a non-zero projection in $\mathcal{M}$, the identity projection I represents the always true proposition (tautology) and the semantic entailment $\alpha \models \beta$ turns out to be given by the partial ordering on $\mathcal{P}(\mathcal{H})$ (resp. on $\mathcal{P}(\mathcal{M})$). As it was seen in Chapter 8, the semantic content of an implication connective is formulated in terms of the implicative criteria (see Section 8.1). So if we consider the case when $\mathcal{F}_1$ and $\mathcal{F}_2$ are represented by the orthomodular lattices $\mathcal{P}(\mathcal{M}_1)$ and $\mathcal{P}(\mathcal{M}_2)$ of two von Neumann subalgebras of the von Neumann algebra $\mathcal{M}$, then the independence definitions can be formulated in terms of the lattice operations of $\mathcal{P}(\mathcal{M})$ as follows.

Definition 11.1 becomes

Definition 11.4 $\mathcal{P}(\mathcal{M}_1), \mathcal{P}(\mathcal{M}_2)$ are *semantically independent* if

$$\text{semantic independence:} \quad A \not\leq B \quad B \not\leq A \quad (11.3)$$

for any elements $0, I \neq A \in \mathcal{P}(\mathcal{M}_1)$ and $0, I \neq B \in \mathcal{P}(\mathcal{M}_2)$.

The lattice theoretic translation of Definition 11.2 is

Definition 11.5 $\mathcal{P}(\mathcal{M}_1), \mathcal{P}(\mathcal{M}_2)$ are called $\Rightarrow$-independent if

$$\Rightarrow\text{-independence:} \quad (A \Rightarrow B) \neq I \text{ and } (B \Rightarrow A) \neq I \quad (11.4)$$

for any $0, I \neq A \in L_1$ and $0, I \neq B \in L_2$.

Finally, the Definition 11.3 of logical independence becomes

Definition 11.6 $\mathcal{P}(\mathcal{M}_1)$ and $\mathcal{P}(\mathcal{M}_2)$ are called logically independent if and only if for any non-zero $A \in \mathcal{P}(\mathcal{M}_1)$ and for any non-zero $B \in \mathcal{P}(\mathcal{M}_2)$ it holds that $A \wedge B \neq 0$.

The content of this last definition can be expressed also by saying that $\mathcal{P}(\mathcal{M}_1), \mathcal{P}(\mathcal{M}_2)$ are logically independent if and only if *any* pair of non-zero propositions (A, B) ($A \in \mathcal{P}(\mathcal{M}_1), B \in \mathcal{P}(\mathcal{M}_2)$) can be jointly true in some state of the (joint) system S. For instance, if A represents the proposition that "The observable $R_1 \in \mathcal{M}_1$ has value r_1 (with probability one)", and B stands for "The observable $R_2 \in \mathcal{M}_2$ has value r_2 (with probability one)", then there is at least one state ϕ of the system S in which the proposition "The observable $R_1 \in \mathcal{M}_1$ has its value r_1 and the observable $R_2 \in \mathcal{M}_2$ has its value r_2 (with probability one)" is true.

Recall (Section 8.1) that one of the implicative criteria, the "law of entailment" required of the $\Rightarrow$ implication the following

(E) $\qquad\qquad$ if $A \leq B$, then $(A \Rightarrow B) = I$

It follows at once that if $\Rightarrow$ satisfies E, then $\Rightarrow$-independence of the lattices $\mathcal{P}(\mathcal{M}_1)$ and $\mathcal{P}(\mathcal{M}_2)$ implies semantic independence of $\mathcal{P}(\mathcal{M}_1), \mathcal{P}(\mathcal{M}_2)$. If $\Rightarrow$ also satisfies the modus ponens

(MP) $\qquad\qquad\qquad A \wedge (A \Rightarrow B) \leq B$

(see Section 8.1), then

$$A \leq B \quad \text{if and only if} \quad (A \Rightarrow B) = I \qquad (11.5)$$

and in this case the semantic independence and $\Rightarrow$-independence of the lattices $\mathcal{P}(\mathcal{M}_1)$ and $\mathcal{P}(\mathcal{M}_2)$ are equivalent.

It is easy to see that the logical independence of $\mathcal{P}(\mathcal{M}_1), \mathcal{P}(\mathcal{M}_2)$ implies semantic independence: If $A \leq B$ for some $A \in \mathcal{P}(\mathcal{M}_1), B \in \mathcal{P}(\mathcal{M}_2)$, then A is orthogonal to $B^\perp$, consequently $A \wedge B^\perp = 0$, and since $B^\perp$ is in $\mathcal{P}(\mathcal{M}_2)$ if B is (because $\mathcal{P}(\mathcal{M}_2)$ is a sublattice), $\mathcal{P}(\mathcal{M}_1), \mathcal{P}(\mathcal{M}_2)$ are not logically independent. But the converse is not true, as the following counterexample shows. Consider the simplest six-element orthomodular lattice:

$$L_6 = \{A, A^\perp, B, B^\perp, 0, I\}$$

with the partial ordering given by

$$0 \leq X \leq I \qquad (X = A, A^\perp, B, B^\perp)$$

If $\Rightarrow$ satisfies the minimal implicative criteria (with respect to this $\leq$), then the two sublattices

$$\begin{aligned} L_1 &= \{0, A, A^\perp, I\} \\ L_2 &= \{0, B, B^\perp, I\} \end{aligned}$$

are $\Rightarrow$-independent (and also semantically independent) but not logically independent.

The next proposition gives a sufficient condition for $\mathcal{P}(\mathcal{M}_1), \mathcal{P}(\mathcal{M}_2)$ that implies that the $\Rightarrow$-independence of $\mathcal{P}(\mathcal{M}_1), \mathcal{P}(\mathcal{M}_2)$ entails logical independence.

Proposition 11.1 *Let us assume that the connective $\Rightarrow$ satisfies the minimal implicative criteria. If the lattices $\mathcal{P}(\mathcal{M}_1)$ and $\mathcal{P}(\mathcal{M}_2)$ are such that the orthomodular lattice generated by any two elements $A \in \mathcal{P}(\mathcal{M}_1)$ and $B \in \mathcal{P}(\mathcal{M}_2)$ is a distributive sublattice in L, then $\Rightarrow$-independence of $\mathcal{P}(\mathcal{M}_1), \mathcal{P}(\mathcal{M}_2)$ implies logical independence of $\mathcal{P}(\mathcal{M}_1), \mathcal{P}(\mathcal{M}_2)$.*

Proof: Assume that $\mathcal{P}(\mathcal{M}_1), \mathcal{P}(\mathcal{M}_2)$ are not logically independent. Then there exist non-trivial projections $A \in \mathcal{P}(\mathcal{M}_1), B \in \mathcal{P}(\mathcal{M}_2)$ such that $A \wedge B = 0$. Using the distributivity assumption we can write then:

$$\begin{aligned} A = A \wedge I &= A \wedge (B \vee B^\perp) \\ = (A \wedge B) \vee (A \wedge B^\perp) &= 0 \vee (A \wedge B^\perp) = A \wedge B^\perp \end{aligned}$$

which imples $A \leq B^\perp$ and so by (11.5) we have $(A \Rightarrow B^\perp) = I$, and consequently $\mathcal{P}(\mathcal{M}_1), \mathcal{P}(\mathcal{M}_2)$ are not $\Rightarrow$-independent. □

In the projection lattice of a von Neumann algebra a sublattice generated by a set of projections is distributive if and only if the projections are pairwise commuting (Propositions 4.16 and 4.17), hence as a particular case of Proposition 11.1 we have:

Proposition 11.2 *Assume that $\mathcal{M}_1, \mathcal{M}_2$ are commuting von Neumann subalgebras of the von Neumann algebra $\mathcal{M}$. If the lattices $\mathcal{P}(\mathcal{M}_1)$ and $\mathcal{P}(\mathcal{M}_2)$ are $\Rightarrow$-independent with respect to any of the quantum conditionals (which means in this case that the lattices are independent with respect to the classical material implication) then $\mathcal{P}(\mathcal{M}_1), \mathcal{P}(\mathcal{M}_2)$ are logically independent.*

Summing up the relations between the three independence notions in the context of quantum logic:

1. Semantic and $\Rightarrow$-independence of $\mathcal{P}(\mathcal{M}_1), \mathcal{P}(\mathcal{M}_2)$ are equivalent if $\Rightarrow$ satisfies the minimal implicative criteria.
2. If $\Rightarrow$ satisfies the minimal implicative criteria, and $\mathcal{P}(\mathcal{M}_1), \mathcal{P}(\mathcal{M}_2)$ are mutually commuting, then the three independence conditions coincide; this is the case in particular when $\mathcal{P}(\mathcal{M})$ is a Boolean algebra (logic of a classical mechanical system).
3. In the non-distributive, truly quantum case, logical independence is strictly stronger than semantic or $\Rightarrow$-independence, even if $\Rightarrow$ satisfies minimal implicative criteria.

It is this stronger notion of logical independence which we now turn to.

11.1.2. LOGICAL AND STATISTICAL INDEPENDENCE

Let us see an example of logically *non-independent* subalgebras first. Let Q and P be the canonically conjugate position and momentum operators of a free particle moving in one dimension: Q and P are defined on (some suitable dense set $\mathcal{D}$ of) the Hilbert space $L^2(\mathbb{R}, \mu)$ by

$$(Pf)(x) = -if'(x) \qquad (Qf)(x) = xf(x) \qquad f \in \mathcal{D}$$

and let P^Q and P^P be the spectral measures of Q and P (see Chapter 2). Let $\mathcal{M}_1 = \mathcal{M}_Q$ and $\mathcal{M}_2 = \mathcal{M}_P$ be the von Neumann algebras generated by the spectral projections of Q and P:

$$\mathcal{M}_Q = \{\mathrm{P}^Q(d) \mid d \text{ real Borel set }\}''$$
$$\mathcal{M}_P = \{\mathrm{P}^P(d) \mid d \text{ real Borel set }\}''$$

Let us also define

$$\mathcal{M}'_Q = \{X \in \mathcal{B}(\mathcal{H}) \mid X\mathrm{P}^Q(d) = \mathrm{P}^Q(d)X \text{ for all real Borel sets } d\}$$
$$\mathcal{M}'_P = \{X \in \mathcal{B}(\mathcal{H}) \mid X\mathrm{P}^P(d) = \mathrm{P}^P(d)X \text{ for all real Borel sets } d\}$$

($\mathcal{B}(\mathcal{H})$ denoting the set of all bounded operators on $\mathcal{H}$, and $\mathcal{S}'$ denoting the first and $\mathcal{S}'' = (\mathcal{S}')'$ denoting the second commutant of a set $\mathcal{S}$ of bounded operators on $\mathcal{H}$.)

By eq. (2.19) the two von Neumann sublattices $\mathcal{P}(\mathcal{M}_Q), \mathcal{P}(\mathcal{M}_P)$ generated by the spectral projections of Q and P respectively are not logically independent.

Note that the lattices $\mathcal{P}(\mathcal{M}'_Q), \mathcal{P}(\mathcal{M}'_P)$ are also not independent logically, since it holds that

$$\mathcal{P}(\mathcal{M}_Q) \subseteq \mathcal{P}(\mathcal{M}'_Q) \quad \mathcal{P}(\mathcal{M}_P) \subseteq \mathcal{P}(\mathcal{M}'_P)$$

and logical non-independence is inherited by super-lattices: if $\mathcal{L}_1$ and $\mathcal{L}_2$ contain sublattices $\mathcal{L}_{10}$ and $\mathcal{L}_{20}$ that are not logically independent, then $\mathcal{L}_1, \mathcal{L}_2$ are also not logically independent. Consequently, the two von Neumann sublattices $\mathcal{P}(\mathcal{M}_Q), \mathcal{P}(\mathcal{M}_P)$ generated by the spectral projections of Q and P respectively are not logically independent. In fact, these lattices are very strongly not independent logically, since they contain an abundance of projections that do not have non-zero greatest lower bound. The logical non-independence of $\mathcal{M}_Q$ and $\mathcal{M}_P$ is due to the strong correlation between complementary quantities, which is a consequence of the well known non-commutativity of Q and P: $[Q,P] = iI$ (on a suitable subset of $L^2(\mathbb{R}, \mu)$).

This leads to the question: are $\mathcal{P}(\mathcal{M}_1)$ and $\mathcal{P}(\mathcal{M}_2)$ logically independent if they mutually commute, i.e. if $AB = BA$ for all $A \in \mathcal{P}(\mathcal{M}_1)$ and for all $B \in \mathcal{P}(\mathcal{M}_1)$? The answer is no, since mutual commutativity of $\mathcal{P}(\mathcal{M}_1), \mathcal{P}(\mathcal{M}_2)$ is not sufficient to imply

$$\mathcal{P}(\mathcal{M}_1) \cap \mathcal{P}(\mathcal{M}_2) = \{0, I\}$$

which is a necessary condition for two sublogics to be logically independent: if there is a non-trivial element A in $\mathcal{P}(\mathcal{M}_1) \cap \mathcal{P}(\mathcal{M}_2)$, then $A^\perp$ also is in $\mathcal{P}(\mathcal{M}_1) \cap \mathcal{P}(\mathcal{M}_2)$, and $A \wedge A^\perp = 0$, thus $\mathcal{P}(\mathcal{M}_1), \mathcal{P}(\mathcal{M}_2)$ cannot be logically independent. (Note that mutual commutativity of $\mathcal{P}(\mathcal{M}_1), \mathcal{P}(\mathcal{M}_2)$ is not only not sufficient to imply logical independence of $\mathcal{P}(\mathcal{M}_1), \mathcal{P}(\mathcal{M}_2)$, it is not even necessary for $\mathcal{P}(\mathcal{M}_1), \mathcal{P}(\mathcal{M}_2)$ to be logically independent, see the example preceeding Problem 2). This leads to the

Problem 11.1 What additional conditions to mutual commutativity imply then logical independence?

A partial answer to this question is contained in the next proposition.

Proposition 11.3 *If $\mathcal{M}_1$ is a factor von Neumann algebra in the von Neumann algebra $\mathcal{M}$, and the lattices $\mathcal{P}(\mathcal{M}_1)$ and $\mathcal{P}(\mathcal{M}_2)$ are commuting, then $\mathcal{P}(\mathcal{M}_1)$ and $\mathcal{P}(\mathcal{M}_2)$ are logically independent.*

This proposition is implied by a result of Murray and von Neumann (Corollary to Theorem III, in [104]): Let $\mathcal{M} \subseteq \mathcal{B}(\mathcal{H})$ be a factor von Neumann algebra, and $A \in \mathcal{M}, B \in \mathcal{M}'$ (recall that $\mathcal{M}'$ denotes the commutant of $\mathcal{M}$). If $AB = 0$, then either $A = 0$ or $B = 0$. In particular, the next proposition, which is just the special case of Proposition 11.3, is implied by this result of Murray and von Neumann; however, because of the significance of the finite von Neumann algebras for quantum logic, a proof of it is given below.

Proposition 11.4 *Let $\mathcal{M}$ be a finite factor von Neumann algebra and $\mathcal{M}_1$ be a subfactor in $\mathcal{M}$. If $\mathcal{M}_2$ commutes with $\mathcal{M}_1$, then $\mathcal{P}(\mathcal{M}_1), \mathcal{P}(\mathcal{M}_2)$ are logically independent.*

Proof: Note first that logical independence is inherited by pairs of subalgebras: If $\mathcal{P}(\mathcal{M}_1), \mathcal{P}(\mathcal{M}_2)$ are logically independent and $\mathcal{M}_{10}, \mathcal{M}_{20}$ are von Neumann subalgebras of $\mathcal{M}_1$ and $\mathcal{M}_2$, respectively, then $\mathcal{P}(\mathcal{M}_{10})$ and $\mathcal{P}(\mathcal{M}_{20})$ also are logically independent. Thus it suffices to show that $\mathcal{M}_1$ and $\mathcal{N} \equiv \mathcal{M}_1' \cap \mathcal{M} \supseteq \mathcal{M}_2$ are logically independent ($\mathcal{M}_1' \cap \mathcal{M}$ denoting the elements in $\mathcal{M}$ that commute with every element in $\mathcal{M}_1$). Let $\mathcal{M}_1 \vee \mathcal{N}$ denote the von Neumann subalgebra in $\mathcal{M}$ generated by $\mathcal{M}_1$ and $\mathcal{N}$, and let τ be the unique, faithful, tracial state on $\mathcal{M}$. The restriction of τ to $\mathcal{M}_1 \vee \mathcal{N}$ is also a faithful tracial state on $\mathcal{M}_1 \vee \mathcal{N}$, and there exists then a unique τ-preserving conditional expectation from $\mathcal{M}_1 \vee \mathcal{N}$ onto the subfactor $\mathcal{M}_1$ of $\mathcal{M}_1 \vee \mathcal{N}$ (Proposition 8.6). By a theorem of Takesaki (Corollary 1 in [157]) τ factorizes on $\mathcal{M}_1, \mathcal{N}$; that is,

$$\tau(AB) = \tau(A)\tau(B) \qquad A \in \mathcal{M}_1, B \in \mathcal{N}$$

Since τ is faithful, it follows that

$$0 \neq \tau(A)\tau(B) = \tau(AB) = \tau(A \wedge B) \quad \text{if} \quad A, B \neq 0$$

which implies $(A \wedge B) \neq 0$ if $A, B \neq 0$. □

Remark: As we have seen (see the table in Section 6.1), there are two types of finite factors: The set of all bounded operators $\mathcal{B}(\mathcal{H}_n)$ on an n-dimensional Hilbert space $\mathcal{H}_n$ (these are the discrete, type I_n finite factors) and the type II_1 factors (continuous case). Proposition 11.3 holds trivially if the algebra $\mathcal{M}_1$ is irreducible in a type II_1 factor $\mathcal{M}$: i.e. if (by definition of irreducibility) the relative commutant $(\mathcal{M}_1' \cap \mathcal{M})$ of $\mathcal{M}_1$ in $\mathcal{M}$ is equal to $\{0, I\}$. By results of the index theory of type II_1 factors, $\mathcal{M}_1$ cannot be an arbitrary subfactor in $\mathcal{M}$ if $\mathcal{M}_1$ is not irreducible in the type II_1 factor $\mathcal{M}$: if the index of $\mathcal{M}_1$ is smaller than 4, then $\mathcal{M}_1$ is irreducible in $\mathcal{M}$ (Corollary 2.2.4 in [78]). However, if $\mathcal{M}$ is the unique hyperfinite type II_1 factor R, then there exist non-irreducible subfactors $\mathcal{M}_1 \subset R$ with index greater than 4 (see [78]). Thus the Proposition 11.3 holds non-trivially in these latter cases.

Remark : Note that the projection lattices of type II_1 (factor) von Neumann algebras, the "continuous geometries", are just the lattices that Birkhoff and von Neumann considered as (irreducible) quantum logics (Chapter 7). As it was seen in Section 6.1 the lattice $\mathcal{P}(\mathcal{M})$ of a type II_1 factor von Neumann algebra $\mathcal{M}$ differs from the Hilbert lattice $\mathcal{P}(\mathcal{H})$ of all projections on the Hilbert space $\mathcal{H}$ in that $\mathcal{P}(\mathcal{M})$ has no atoms and it is modular, unlike $\mathcal{P}(\mathcal{H})$, which is not modular if $\mathcal{H}$ is infinite dimensional, and which is atomic irrespective of the dimension of $\mathcal{H}$. If $\mathcal{P}(\mathcal{M}_1), \mathcal{P}(\mathcal{M}_2)$ are logically independent, and both $\mathcal{P}(\mathcal{M}_1)$ and $\mathcal{P}(\mathcal{M}_2)$ contain a non-trivial element, then neither $\mathcal{P}(\mathcal{M}_1)$ nor $\mathcal{P}(\mathcal{M}_2)$ can contain any atom of $\mathcal{P}(\mathcal{M})$. Because of the symmetrical role of $\mathcal{P}(\mathcal{M}_1)$ and $\mathcal{P}(\mathcal{M}_2)$ to see this it is enough to show that if $0, I \neq A_1 \in \mathcal{P}(\mathcal{M}_1)$ is a non-trivial element and $A_0 \in \mathcal{P}(\mathcal{M}_2)$ is an atom in $\mathcal{P}(\mathcal{M})$, then $\mathcal{P}(\mathcal{M}_1), \mathcal{P}(\mathcal{M}_2)$ are not logically independent. It holds that

$$A_1 \wedge A_0 \leq A_0 \tag{11.6}$$

and since A_0 is an atom, it follows from (11.6) that either $A_1 \wedge A_0 = 0$, which means that $\mathcal{P}(\mathcal{M}_1), \mathcal{P}(\mathcal{M}_2)$ are not logically independent, or $A_1 \wedge A_0 = A_0$, which implies $A_1 \geq A_0$, and then for the non-zero element $A_1^\perp$ we have $A_1^\perp \wedge A_0 = 0$, thus, again, $\mathcal{P}(\mathcal{M}_1), \mathcal{P}(\mathcal{M}_2)$ are not logically independent. Neither $\mathcal{P}(\mathcal{M}_1)$ nor $\mathcal{P}(\mathcal{M}_2)$ contains an atom of $\mathcal{P}(\mathcal{M})$ if $\mathcal{M}$ is a type II_1 factor, since there are no atoms at all in $\mathcal{P}(\mathcal{M})$ in this case; it is not true, however, that $\mathcal{P}(\mathcal{M}_1), \mathcal{P}(\mathcal{M}_2)$ cannot be logically independent non-trivially, if $\mathcal{P}(\mathcal{M})$ is an atomic lattice (see examples below).

Next we relate logical independence to the statistical independence condition known as C^*-independence. Recall that C^*-independence of the C^*-subalgebras $\mathcal{A}, \mathcal{B}$ of the C^*-algebra $\mathcal{C}$ means that for any state ϕ_1 on $\mathcal{A}_1$ and for any state ϕ_2 on $\mathcal{A}_2$ there is a state ϕ on $\mathcal{C}$ such that $\phi(A) = \phi_1(A)$ and $\phi(B) = \phi_2(B)$ ($A \in \mathcal{A}_1, B \in \mathcal{A}_2$) (Definition 10.3).

Clearly, if $\mathcal{M}_1, \mathcal{M}_2$ are mutually commuting C^*-independent von Neumann algebras, then $\mathcal{P}(\mathcal{M}_1), \mathcal{P}(\mathcal{M}_2)$ are logically independent by the Schlieder-Ross Theorem (Proposition 10.5), since $A \wedge B = AB$ for commuting A and B. The converse also is true, for assume that $\mathcal{M}_1, \mathcal{M}_2$ are commuting and not C^*-independent. Then $XY = 0$ for some $0 \neq X \in \mathcal{M}_1$

and $0 \neq Y \in \mathcal{M}_2$, and then the non-zero range projection (left support) $s_l(Y)$ of Y (which lies in $\mathcal{M}_2$) and the non-zero range projection of X^* (which belongs to $\mathcal{M}_1$, and which is the right support $s_r(X)$ of X) are related as

$$s_l(Y) \subseteq \ker(X) = I - \text{range}(X^*)$$
$$= s_r(X)^\perp$$

and it follows that $s_l(Y) \leq s_r(X)^\perp$, that is $s_l(Y)$ and $s_r(X)$ are orthogonal. Consequently $s_l(Y) \wedge s_r(X) = 0$, and so $\mathcal{P}(\mathcal{M}_1), \mathcal{P}(\mathcal{M}_2)$ are not logically independent. So we have

Proposition 11.5 *If $\mathcal{M}_1, \mathcal{M}_2$ is a commuting pair of von Neumann subalgebras of the von Neumann algebra $\mathcal{M}$, then the pair $(\mathcal{P}(\mathcal{M}_1), \mathcal{P}(\mathcal{M}_2))$ is logically independent if and only if $\mathcal{M}_1, \mathcal{M}_2$ are C^*-independent.*

In particular, the logics $\mathcal{P}(\mathcal{M}(V_1)), \mathcal{P}(\mathcal{M}((V_2))$ associated to the von Neumann algebras $\mathcal{M}(V_1), \mathcal{M}(V_2)$ of local observables localized in spacelike separated wedge and double cone regions V_1, V_2 in the Minkowski space in the sense of algebraic quantum field theory (Chapter 10) are logically independent, since the algebras $\mathcal{M}(V_1), \mathcal{M}(V_2)$ commute by the axiom of microcausality and they also are known to be C^*-independent. The algebras $\mathcal{M}(V_1), \mathcal{M}(V_2)$ being type **III** typically, the von Neumann lattice $\mathcal{P}(\mathcal{M}(V_1 \cup V_2))$ does not contain atoms.

Another example of C^*-independent pair of algebras is the pair

$$(M_n \otimes I, I \otimes M_n)$$

more generally, the pair

$$(\mathcal{B}(\mathcal{H}) \overline{\otimes} I, I \overline{\otimes} \mathcal{B}(\mathcal{H}))$$

in $\mathcal{B}(\mathcal{H}) \overline{\otimes} \mathcal{B}(\mathcal{H})$. Here M_n is the algebra of complex n-by-n matrices, and the two algebras $(M_n \otimes I)$ and $(I \otimes M_n)$ is considered as subalgebras of the matrix algebra

$$M_n \otimes M_n = M_{n^2}$$

The case $n = 2$ is the well known Bohm-Bell joint system of (the spin part of) two spin half particles. Since $(M_n \otimes I)$ and $(I \otimes M_n)$ commute,

$$\mathcal{P}(M_n \otimes I), \mathcal{P}(I \otimes M_n)$$

form a logically independent pair of quantum logic, which is immediately clear also without invoking Proposition 11.4: if $(A \otimes I)$ and $(I \otimes B)$ are projections in $(M_n \otimes I)$ and $(I \otimes M_n)$, respectively, then

$$(A \otimes I) \wedge (I \otimes B) = (A \otimes B) \neq 0$$

In this example $\mathcal{P}(\mathcal{M}_{n^2})$ is an atomic lattice but neither $\mathcal{P}(\mathcal{M}_n \otimes I)$ nor $\mathcal{P}(I \otimes \mathcal{M}_n)$ contains an atom of $\mathcal{P}(\mathcal{M}_{n^2})$ (cf. the second Remark after Proposition 11.4). Note that $\mathcal{M}_n \otimes I$ is a subfactor in $\mathcal{M}_{n^2}$, thus the logical independence of the pair $(\mathcal{M}_n \otimes I), (I \otimes \mathcal{M}_n)$ follows from Proposition 11.3, too.

It is known [151] that there are C^*-independent pairs of algebras that do not commute. Consider now the five dimensional Euclidean space $\mathbb{R}^5$ (denote the axis in $\mathbb{R}^5$ by X_i, $i = 1, \ldots 5$), let A be the X_1, X_2 plane and B_0 the plane at angle $\pi/5$ to A which contains the X_1 axis, and let $B = B_0 + X_5$. Then the two lattices $(A, A^\perp, 0, I)$ and $(B, B^\perp, 0, I)$ do not commute and are logically independent. (This is because $(A \wedge B)$ is equal to the X_1 axis, the X_4 axis is in $A^\perp \wedge B^\perp$, X_5 is in both $A^\perp \wedge B$ and $A \wedge B^\perp$.) So one is led to the following two problems:

Problem 11.2 What is the relation of the notion of C^*-independence to logical independence in general, i.e. in the case where $\mathcal{P}(\mathcal{M}_1), \mathcal{P}(\mathcal{M}_2)$ are not supposed to be mutually commuting?

Problem 11.3 What additional conditions to the logical independence of the pair $(\mathcal{P}(\mathcal{M}_1), \mathcal{P}(\mathcal{M}_2))$ imply that the two von Neumann lattices $\mathcal{P}(\mathcal{M}_1)$ and $\mathcal{P}(\mathcal{M}_2)$ commute?

Note that while it is true that the Schlieder-Roos theorem remains valid without the assumption of mutual commutativity of the algebras involved, this fact together with Proposition 11.5 does not give an answer to the question in Problem 11.2 because if $\mathcal{M}_1$ does not commute with $\mathcal{M}_2$, then, if $A \in \mathcal{M}_1$ and $B \in \mathcal{M}_2$ are projections, then $AB \neq A \wedge B$ (AB is not a projection, not even selfadjoint, if A and B do not commute).

Partial answer to Problem 11.2 is given by the next

Proposition 11.6 *Let $\mathcal{M}$ be a finite dimensional full matrix algebra and $\mathcal{M}_1, \mathcal{M}_2$ be two, not necessarily commuting von Neumann subalgebras of $\mathcal{M}$. If the pair $(\mathcal{M}_1, \mathcal{M}_2)$ is C^*-independent, then $\mathcal{P}(\mathcal{M}_1), \mathcal{P}(\mathcal{M}_2)$ are logically independent.*

Proof: The assertion of the proposition is a consequence of the fact that the projection lattice of a finite dimensional full matrix algebra is a so-called Jauch-Piron lattice: The lattice $\mathcal{P}(\mathcal{M})$ of an arbitrary von Neumann algebra $\mathcal{M}$ is a *Jauch-Piron lattice* by definition if *every* state on $\mathcal{M}$ is a Jauch-Piron state. (Definition 3.23). A state ϕ is Jauch-Piron (Definition 3.22) if it satisfies the following condition: whenever $A, B \in \mathcal{P}(\mathcal{M})$ are such that
$$\phi(A) = \phi(B) = 0$$
it holds that
$$\phi(A \vee B) = 0$$

Equivalently:
$$\phi(A) = \phi(B) = 1$$
implies
$$\phi(A \wedge B) = 1$$
Since for any element X in a C^*-algebra there is a state ϕ such that
$$\phi(XX^*) = \| X \|^2$$
for any $A \in \mathcal{M}_1$ and $B \in \mathcal{M}_2$, there exist states ϕ_1 and ϕ_2 on $\mathcal{M}_1$ and $\mathcal{M}_2$, respectively, such that $\phi_1(A) = 1$ and $\phi_2(B) = 1$. By the C^*-independence of the pair $\mathcal{P}(\mathcal{M}_1), \mathcal{P}(\mathcal{M}_2)$ there is a joint extension ϕ of ϕ_1 and ϕ_2, and since ϕ is Jauch-Piron, it follows that
$$\phi(A \wedge B) = 1$$
hence $A \wedge B \neq 0$. □

Remark: The above argument shows that somewhat more is true than what is stated in Proposition 11.6: If $\mathcal{M}$ is such that $\mathcal{P}(\mathcal{M})$ is a Jauch-Piron lattice, then two sub-von Neumann lattices $\mathcal{P}(\mathcal{M}_1), \mathcal{P}(\mathcal{M}_2)$ are logically independent whenever $\mathcal{M}_1, \mathcal{M}_2$ are C^*-independent. A complete characterization of the Jauch-Piron property is known in the von Neumann algebra category: If $\mathcal{M}$ does not contain an I_2 direct summand, then $\mathcal{P}(\mathcal{M})$ is Jauch-Piron if and only if it is the direct sum of a commutative algebra and finitely many finite dimensional factors [59]. For all such von Neumann algebras Proposition 11.6 applies. Since the projection lattice of a finite von Neumann algebra is not necessarily Jauch-Piron, one is led to the following

Problem 11.4 Does C^*-independence of the von Neumann algebras $\mathcal{M}_1$ and $\mathcal{M}_2$ imply logical independence of $\mathcal{P}(\mathcal{M}_1)$ and $\mathcal{P}(\mathcal{M}_2)$, if $(\mathcal{M}_1, \mathcal{M}_2)$ is a not necessarily commuting pair of von Neumann subalgebras of a finite von Neumann algebra $\mathcal{M}$?

The two other statistical independence conditions that are relevant for logical independence are: W^*-independence (Definition 10.4) and strict locality (Definition 10.6).

It is easy to see that if $\mathcal{M}_1, \mathcal{M}_2$ are commuting, then strict locality of $\mathcal{M}_1, \mathcal{M}_2$ implies C^*-independence of $\mathcal{M}_1, \mathcal{M}_2$: Assume that $\mathcal{M}$ acts on the Hilbert space $\mathcal{H}$, and let $0 \neq A \in \mathcal{P}(\mathcal{M}_1)$ and $0 \neq B \in \mathcal{P}(\mathcal{M}_2)$ be arbitrary projections. Let ϕ_2 be a vector state on $\mathcal{M}_2$ given by a vector in B. By strict locality there exists a normal state ϕ on $\mathcal{M}$ such that $\phi(A) = 1$ and $\phi(B) = \phi_2(B) = 1$. Since ϕ is normal, it is given by a density matrix w, which can be written in its spectral resolution as $w = \sum_i \lambda_i P_i$. Since $\phi(A) = \phi(B) = 1$ and $\sum_i \lambda_i = 1$, it follows that every projection P_i is contained in both A and in B, so if $AB = A \wedge B$ were equal to 0, then w would be zero and ϕ would be zero, which contradicts $\phi(A) = \phi(B) = 1$. Thus $AB \neq 0$ for any two projections. By the reasoning preceding Proposition 11.5 one concludes that $Q_1 Q_2 \neq 0$ for any $0 \neq Q_1 \in \mathcal{M}_1$ and for any $0 \neq Q_2 \in \mathcal{M}_2$, and so by the Schlieder-Roos theorem (Proposition 10.5) $\mathcal{M}_1, \mathcal{M}_2$ are C^*-independent. It follows then that

Proposition 11.7 $\mathcal{P}(\mathcal{M}_1), \mathcal{P}(\mathcal{M}_2)$ *are logically independent if* $(\mathcal{M}_1, \mathcal{M}_2)$ *is a strictly local pair of commuting von Neumann algebras.*

While it is known that strict locality implies C^*-independence even if $\mathcal{M}_1, \mathcal{M}_2$ do not commute (Corollary 10 in [51], it does not follow from this alone that if $\mathcal{M}_1$ and $\mathcal{M}_2$ are not commuting and strictly local, then they are logically independent, again because if A and B do not commute, then $AB \neq A \wedge B$. Thus the following problem arises:

Problem 11.5 What is the relation of logical independence to strict locality in the general case, i.e. if $\mathcal{M}_1, \mathcal{M}_2$ are not assumed to be mutually commuting?

The answer to this problem is given by the next proposition:

Proposition 11.8 *Let* $\mathcal{M}$ *be an arbitrary von Neumann algebra and* $\mathcal{M}_1$ *and* $\mathcal{M}_2$ *be two, not necessarily commuting von Neumann subalgebras of* $\mathcal{M}$. *Then the pair* $(\mathcal{P}(\mathcal{M}_1), \mathcal{P}(\mathcal{M}_2))$ *is logically independent if* $(\mathcal{M}_1, \mathcal{M}_2)$ *is a* W^*-*independent or strictly local pair.*

Proof: The proposition is an immediate consequence of the fact that every normal state on a von Neumann algebra is a Jauch-Piron state. (The proof of this fact is the same as the proof of Proposition 4.6 .) Assume that $\mathcal{M}$ acts on the Hilbert space $\mathcal{H}$, and let A, B be two non-zero projections in $\mathcal{M}_1$ and $\mathcal{M}_2$ respectively. Then there are non-zero unit vectors $\xi \in A$ and $\eta \in B$, and for the corresponding vector (hence normal) states ω_ξ and ω_η we have $\omega_\xi(A) = 1$ and $\omega_\eta(B) = 1$. If $\mathcal{M}_1, \mathcal{M}_2$ are W^*-independent, then there exists a normal extension ϕ of ω_ξ and ω_η, and since ϕ is Jauch-Piron, it holds that $\phi(A \wedge B) = 1$, hence $A \wedge B \neq 0$. Let $(\mathcal{M}_1, \mathcal{M}_2)$ be a strictly local pair and let A, B be as before. If we now take the state ω_ξ on $\mathcal{M}_1$, then strict locality ensures the existence of a normal state ϕ on $\mathcal{M}$ such that $\phi(B) = 1$, and again the Jauch-Piron property of ϕ implies $A \wedge B \neq 0$. □

11.2. Counterfactual probabilistic independence

The chief aim of this section is to analyze the problem of causal dependence between spacelike separated events – as these events are described in the operator algebraic framework of relativistic quantum field theory. By a "counterfactual analysis" is meant elaborating in Stalnaker's possible world semantics of counterfactuals Lewis' idea of "counterfactual chancy causation" as this is proposed in [93].

11.2.1. CONCEPT OF COUNTERFACTUAL PROBABILISTIC INDEPENDENCE

The idea of the counterfactual chancy causation is due to David Lewis. According to his analysis, an event F causes the event E counterfactually, if the probability that E happens would be greater if F occured than it would be if F did not occur. This idea can be formulated somewhat more explicitly as follows

Definition 11.7 We say that the event E is caused by the event F in the sense of counterfactual probabilistic causation, if the following two counterfactuals with probabilistic consequents are true:

$$O(F) \to \mathrm{Prob}(O(E)) = r \qquad -(O(F)) \to \mathrm{Prob}(O(E)) = s \qquad (11.7)$$

for some real numbers r, s such that $r > s$, where $O(F)$ (resp. $-O(F)$) is the proposition stating that the event F occurs (resp. does not occur), and Prob() is the chance function (probability) that E occurs.

Lewis points out that chances, and probabilities, are time dependent; consequently, "The actual chance of E [i.e. $\mathrm{Prob}(O(E))$ in the first counterfactual in (11.7)] is to be its chance at the time immediately after F; and the counterfactual is to concern chance at that same time." ([93], p. 176-177) Butterfield, too, requires the Prob() function to be "... the chance function just after the time (specelike hypersurface) that F occurs or does not occur, as the case may be." ([36] p.) But just *when* precisely is "the time immediately after F"? It *must* be a well-defined time if the probabilities are indeed time dependent, otherwise $\mathrm{Prob}(E)$ in the above definition of counterfactual causal dependence, thus Definition 11.7 itself becomes indefinite. It is worrysome then that "time just after the time an event occurs" does not seem to be well defined, since there is no such thing as the "time immediately after a given time".

This ambiguity in connection with Lewis' (and Butterfield's) definition of counterfactual causal dependence can (and will below) be circumvented by a consequent application of the point of view of relativity theory, where there is no "time" (as a frame independent parameter); rather, time is viewed as being encoded in the location of the event in the spacetime. This is the case also in ARQFT, to which the above Definition 11.7 will be applied here. Furthermore, the exact "time" of occurrence of an event in ARQFT is not well-defined for an additional reason: ARQFT is a "smeared" theory:

the physically meaningful entities are considered in ARQFT not as being specified at single spacetime points but as belonging to (open, bounded) *regions* in spacetime. Therefore, the events in ARQFT, too, are specified only up to the precision of occurring in certain spacetime *regions*, they can not be considered as being allocated to a definite single point in the spacetime. Thus "the time" of the occurrence of F and E does not make sense if F and E are events in ARQFT (not even if a frame is fixed); on the other hand, their "smeared" location in time is an organic part of their definition as events belonging to a spacetime region. This explains how "time" features in the consideratios to follow although it is not mentioned explicitly.

The meaning of the counterfactual notion of chancy causation as given by Definition 11.7 is not fixed until one gives a prescription for the evaluation of the truth of the two counterfactuals (11.7). Here the idea is taken seriously that counterfactuals are to be evaluated according to possible world semantics. In particular, we use Stalnaker's semantics. Recall (see Section 8.1) that in the Stalnaker's semantics one assumes that there exists a set W of all physically possible worlds, and for every proposition O a function $S_O: W \to W$, the so-called Stalnaker selection function is given, which assigns to every proposition O and every world w the possible world $S_O(w)$ in W which is 'most similar' to w and in which O is true. The counterfactual $O \to R$ is then defined to be true at world w iff R is true at world $S_O(w)$. Under Stalnaker's semantics Definition 11.7 takes on the following form:

Definition 11.8 We say that there exists *counterfactual probabilistic causal dependence* between the events F and E *at world* w iff the two counterfactuals

$$O(F) \to Prob(O(E)) = r \qquad -(O(F)) \to Prob(O(E)) = s \quad (11.8)$$

are true in the sense of Stalnaker semantic for some $r > s$ with some selection function S_X.

The above definition gives the condition for counterfactual probabilistic dependence to exist between two events *at a given world*. The counterfactual probabilistic dependence (in the sense of the above definition) is thus not a property of the pair of the events but a relation between a pair of events *and* a given possible world. This possible-world-dependent notion of counterfactual dependence turns out to be too week to serve as a basis to define (by negation) a notion of counterfactual probabilistic independence between local observable algebras that fits naturally into the hierarchy of other (statistical) independence conditions definable for a pair of algebras of observables. The next definition yields a stronger notion of counterfactual probabilistic dependence that makes the counterfactual probabilistic dependence a property of a pair of events.

Definition 11.9 The event E is said to depend on F in the counterfactual probabilistic sense if and only if for *every* possible world w the above two counterfactuals (11.8) are true (in the sense of Stalnaker's semantics with some selection function) with r and s (possibly depending on w) such that $r \geq s$ (with $r = s$ only in the trivial case $r = s = 0$).

Definition 11.9 is still not a mathematical definition; to obtain one we wish to specify it in ARQFT. This is done in the next section.

11.2.2. COUNTERFACTUAL PROBABILISTIC INDEPENDENCE IN QUANTUM FIELD THEORY

We wish to apply the above sketched theory of counterfactual chancy causation to ARQFT. To do that the following assumptions/identifications are made:

1. The possible local events F, E, Z, etc. in a spacetime region V and the corresponding propositions $O(F), O(E), O(Z)$ spelling out the fact that F, E and Z are the case are identified with the projections in the von Neumann algebra $\mathcal{N}(V)$ pertaining to V.
2. A possible world is represented by a vector state given by a vector $\xi \in \mathcal{H}$ over the quasilocal algebra $\mathcal{N}$. Thus the the possible worlds are in 1-1 correspondence with the vectors in $\mathcal{H}$.
3. Probabilities are determined by (vector) states and (vector) states by probabilities, i.e. given the probabilities $Prob(Z)$ of all events Z in a local algebra $\mathcal{N}(V)$ there is a vector ξ in $\mathcal{H}$ such that

$$Prob(Z) = \frac{\langle \xi, Z\xi \rangle}{\|\xi\|} \qquad (11.9)$$

and each state ξ defines the probabilities of all events via the formula (11.9). If $\| \xi \| = 0$, i.e. if the state is given by the zero vector in $\mathcal{H}$, then the probability of every event is defined to be zero.

4. For a given Z and possible world ξ the Stalnaker selection function is given by

$$S_Z(\xi) \equiv Z\xi. \qquad (11.10)$$

This choice of the selection function means in particular that the world ξ is a Z-world iff $Prob(Z) = 1$ at world ξ (except for the world $\xi = 0$, which is a Z-world for every Z by definition and at which, by definition, the probability of every event is zero).

Since it is the above identifications that give content to the definition of counterfactual probabilistic (in)dependence, a few words in their support are very much in order.

The identification 1. is standard, it is the backbone of quantum logic. And so is the definition of Z−world in 4.: The common interpretation of a quantum proposition is that it is represented by the (closed) subspace (equivalently: by the projection on the subspace) that is spanned by those (vector) states in which the proposition Z is true. The only non-standard feature in 1. is that the set of events is now not (isomorphic to) the lattice $\mathcal{P}(\mathcal{H})$ of all projections on a Hilbert space, since the local algebras in quantum field theory are typically type **III** algebras. The von Neumann lattices of the local algebras thus have properties different from $\mathcal{P}(\mathcal{H})$. For instance they are not atomic, and every projection in the von Neumann lattic of a type **III** algebra are infinite. (This latter fact will be used below to show that local algebras belonging to spacetime regions that are causally dependent are *not* independent in the counterfactual probabilistic sense.)

The identification in 3. means assuming that ARQFT is capable of saying (through the vector states) all physically meaningful probability statements.

The identification 2. is a natural consequence of how ARQFT is assumed to describe physical reality: The physical, intuitive picture behind the formalism of ARQFT is that a concrete net of local algebras fixes the locally observable physical quantities, and a "real", actual physical situation in which these observable quantities have a definite (expectation) value is obtained once we fix the state of the quantum(field) system.

The identification 4. is crucial since the meaning of the counterfactual is encoded in the selection function. Note that specifying the state space of a quantum system as the set of possible worlds in connection with a counterfactual analysis is not uncommon: As we have seen in Section 8.1 one chooses the set of (pure) states as the set of possible worlds and proves that the (Mittelstaedt) quantum conditional is a counterfactual conditional in the sense of Stalnaker semantics if the selection function is exactly the one given by (11.10) (see Proposition 8.5). Given an event Z and a vector ξ, this selection function picks the vector in Z that is closest in Hilbert space norm to ξ. So the similarity between possible worlds is measured in the Hilbert space norm.

Having the identifications one can transform Definition 11.9 into the following

Definition 11.10 Let $E, F \in \mathcal{M}$ be two projections. We say that E *depends on F in the counterfactual probabilistic sense* iff for any $\xi \in \mathcal{H}$ the following two counterfactuals are true in the sense of Stalnaker's semantics

INDEPENDENCE IN QUANTUM LOGIC APPROACH   209

with (11.10) as the selection function

$$F \to \quad Prob(E) = r \qquad F^\perp \to \quad Prob(E) = s \qquad (11.11)$$

for some real numbers r, s (in general, depending on ξ) such that $r \geq s$ (with $r = s$ only in the trivial case $r = s = 0$). We say that there is *counterfactual probabilistic causal dependence* between the two local algebras $\mathcal{N}(V_1), \mathcal{N}(V_2)$ iff there exist two non-trivial projections $A \in \mathcal{N}(V_1), B \in \mathcal{N}(V_2)$ such that either A depends on B or B depends on A (in the counterfactual probabilistic sense specified above). Accordingly, we say that the two algebras are *free of counterfactual probabilistic causal dependence* iff there is no counterfactual probabilistic dependence between them. ARQFT is said to be free of counterfactual probabilistic *superluminal* causation iff any two local algebras $\mathcal{N}(V_1)$ and $\mathcal{N}(V_2)$ belonging to *spacelike separated* regions V_1 and V_2 are free of counterfactual causal dependence.

The question to ask now is this:

Problem 11.6 Is ARQFT free of counterfactual superluminal causal dependence in the sense of the above definition?

As mentioned we are not able to give a general "no" or "yes" answer to this question. What is shown below is that if the two spacelike separated local algebras are statistically independent in the sense of C^*-independence, then they are free of counterfactual causal dependence.

Proposition 11.9 *If $(\mathcal{N}(V_1), \mathcal{N}(V_2))$ is a C^*-independent pair of local algebras belonging to spacelike separated regions V_1, V_2, then the pair is free of counterfactual causal dependence.*

Proof: We must show that for *any* pair of non-trivial projections $A \in \mathcal{N}(V_1), B \in \mathcal{N}(V_2)$ there exists a vector $\xi \in \mathcal{H}$ such that $Prob(B)$ at the world $S_A(\xi) = A\xi$ is equal to r and $Prob(B)$ at the world $S_{A^\perp}(\xi) = A^\perp \xi$ is equal to s with some s and r such that $s > r$. Since $Prob(B)$ at world $S_A(\xi)$ is just

$$\frac{\langle A\xi, BA\xi \rangle}{\| A\xi \|}$$

and similarly $Prob(B)$ at world $S_{A^\perp}(\xi)$ is just

$$\frac{\langle A^\perp \xi, BA^\perp \xi \rangle}{\| A^\perp \xi \|}$$

we must find a ξ such that for some $s > r$ the following hold:

$$\frac{\langle \xi, ABA\xi \rangle}{\| A\xi \|} = r \qquad (11.12)$$

$$\frac{\langle \xi, A^\perp B A^\perp \xi \rangle}{\| A^\perp \xi \|} = s \qquad (11.13)$$

Since $\mathcal{N}(V_1)$ commutes with $\mathcal{N}(V_2)$ by the axiom of microcausality and since $\mathcal{N}(V_1), \mathcal{N}(V_2)$ are C^*-independent by assumption we have by the Schlieder-Roos theorem (Proposition 10.5) that

$$0 \neq A^\perp B = A^\perp \wedge B$$

Hence there exists a non-zero vector ξ that belongs to both $A^\perp$ and B, therefore for this vector (11.12) and (11.13) hold with $r = 0$ and $s = 1$. □

This little argument shows that somewhat more is true than what is stated in the above Proposition: If the two von Neumann algebras $\mathcal{N}, \mathcal{M}$ are such that for any $0 \neq A \in \mathcal{N}$ and $0 \neq B \in \mathcal{M}$ we have $A \wedge B \neq 0$, then the two algebras are free of counterfactual probabilistic dependence. Recall that the pair $(\mathcal{N}, \mathcal{M})$ satisfying the condition

$$A \wedge B \neq 0 \qquad 0 \neq A \in \mathcal{N}, 0 \neq B \in \mathcal{M}$$

is called a logically independent pair (Definition 11.6), and it is known that for two von Neumann algebras $\mathcal{N}$ and $\mathcal{M}$ to be logically independent mutual commutativity of $\mathcal{N}$ and $\mathcal{M}$ is not needed: If the pair is W^*-independent, or if both are finite dimensional matrix algebras and are C^*-independent, then the pair is logically independent, whether or not $\mathcal{N}$ commutes with $\mathcal{M}$ (Propositions 11.6 and 11.7). Thus mutual commutativity is not necessary for counterfactual probabilistic independence.

The two conditions (11.12) and (11.13) show the content of the counterfactual probabilistic dependence under the present specification: B depends on A if and only if conditionalizing *any* probability of B by A using the Lüders rule, the A–conditional probability of B is never smaller than the similarly obtained $A^\perp$–conditional probability of B. The necessary and sufficient condition for counterfactual probabilistic dependence to exist between two algebras is therefore that the two algebras contain non-trivial projections $A \in \mathcal{N}(V_1)$ and $B \in \mathcal{N}(V_2)$ such that $B \leq A$. This is because in this case $A^\perp \leq B^\perp$ and so for any ξ:

$$\frac{\langle \xi, ABA\xi \rangle}{\| A\xi \|} \geq \frac{\langle \xi, A^\perp B A^\perp \xi \rangle}{\| A^\perp \xi \|} = 0 \qquad (11.14)$$

In harmony with Proposition 11.9, if $\mathcal{N}(V_1), \mathcal{N}(V_2)$ are C^*-independent then there are no projections $A \in \mathcal{N}(V_1), B \in \mathcal{N}(V_2)$ such that $A \leq B$ or $B \leq A$: If $A \leq B$ then there are two states ϕ_1 and ϕ_2 on $\mathcal{N}(V_1)$ and $\mathcal{N}(V_2)$ respectively such that $\phi_1(A) = 1$ and $\phi_2(B) = 0$, and these two states cannot have a joint extension.

It is known that algebras belonging to spacelike separated double cones and spacelike separated wedges are C^*-independent. Therefore, the algebras belonging to these spacetime regions are free of counterfactual probabilistic causal dependence. On the other hand, Bell's inequality *is* (maximally) violated (in some states) for observables localized in strictly spacelike separated double cones and wedges; what is more, Bell's inequality is maximally violated in *every normal* (hence also *every vector*) state for algebras belonging to *complementary* wedges and spacelike separated *tangent* double cones. Consequently we have the following

Proposition 11.10 *Violation of Bell's inequality in quantum field theory for observables belonging to spacelike separated spacetime regions does not entail probabilistic counterfactual superluminal causation in quantum field theory in the sense specified in Definition 11.10.*

While one expects spacelike separated C^*-independent local algebras to be independent in the probabilistic counterfactual causal sense, one does not expect the same for algebras that are not located in spacelike separated regions. Let C be a bounded open region in the Minkowski spacetime such that $C^- \setminus C$ is large enough to contain an open bounded region V (recall that C^- is the the causal hull of C). (There is such a C: for instance one can choose a "cylinder" of height $2a$ and width $2b$:

$$C \equiv \{x \in M \ \mid \ |x^0| < a, |x| < b\}$$

C^- is then the "diamond" determined by the cylinder.) Since the region V is causally dependent on C, one expects the two local algebras $\mathcal{N}(V)$ and $\mathcal{N}(C)$ not to be free of counterfactual probabilistic causal dependence. This is indeed so: Take a non-trivial projection B in $\mathcal{N}(V)$. The local algebras being type **III**, they contain only infinite projections and so by the Halving Lemma ([83] p. 412., or [153] p. 23.) there exist an infinite projection $A \in \mathcal{N}(V)$ such that both A and $B - A$ are infinite projections and

$$B \sim A \sim (B - A)$$

($\sim$ being the Murray - von Neumann equivalence relation on the projection lattice, see Section 6.2). By isotony and local primitive causality we have

$$\mathcal{N}(V) \subseteq \mathcal{N}(C^-) = \mathcal{N}(C)$$

so the projection B can be considered as belonging to $\mathcal{N}(C)$, hence we have found two non-trivial projections $A \in \mathcal{N}(V), B \in \mathcal{N}(C)$ such that $A < B$, consequently the two algebras $\mathcal{N}(V)$ and $\mathcal{N}(C)$ are not free of counterfactual causal dependence. This means in particular that Definition 11.10 does distinguish pairs of algebras that belong to causally dependent regions from those that are expected to be independent.

Since the pair of algebras $\mathcal{M}_2 \otimes I, I \otimes \mathcal{M}_2$ ($\mathcal{M}_2$ being the algebra of complex two-by-two matrices) are C^*-independent, it follows *formally* form Proposition 11.9 that, on the present specification of counterfactual probabilistic causation, there is no counterfactual causation involved in the Bohm-Bell system either. This is in contrast to the conclusion of [36] that there *is* superluminal causal dependence (in counterfactual probabilistic sense) between the events in the two wings in the Bohm-Bell system. This contrast is only formal, however, for two reasons: First, the present analysis *does not* apply to the Bohm-Bell system at all, since the system $\mathcal{M}_2 \otimes I, I \otimes \mathcal{M}_2$ is not relativistic in the sense that $\mathcal{M}_2 \otimes I$ and $I \otimes \mathcal{M}_2$ cannot be considered as $\mathcal{N}(V_1)$ and $\mathcal{N}(V_2)$ for some (open, bounded) regions V_1, V_2 in the Minkowski spacetime (because the local algebras in ARQFT are not type I von Neumann algebras). Second, in the present paper's specification the notion of counterfactual probabilistic dependence between events (algebras) differs from and is stronger than Butterfield's definition of counterfactual dependence. In fact, because of the universal quantifier over the possible worlds in Definition 11.9 and in Definition 11.10, one may consider the present specification of counterfactual probabilistic dependence to be too strong. That it is strong indeed is also reflected by the fact that the violation of this dependence, hence counterfactual probabilistic *in*dependence of commuting local algebras, is implied by C^*-independence, which is the weakest statistical independence condition in the hierarchy of the statistical independence notions. So one should further look for possible weakenings of the definitions given in this section. There is a natural constraint that any possible weakening should meet, however: the resulting notion of counterfactual probabilistic (in)dependence of algebras should distinguish local observable algebras that belong to causally non-independent regions from those that are associated with spacelike separated ones.

We conclude with mentioning that there exist other independence concepts not mentioned here; such as Stochastic Einstein Locality (SEL) [67], [126], its strengthening called Stochastic Haag Locality [103] and of different versions [38] of these notions. SEL requires, roughly, that, if for any event E, the behavior of the physical system is fixed throughout the backward light cone of E, then the probability of E is already determined uniquely. As it turns out ARQFT *does* satisfy the SEL property. Thus, if one considers the SEL property as an appropriate prohibition of superluminal causation, then one must conclude that ARQFT is free of superluminal causation also in the sense of SEL – despite violations of Bell's inequalities.

It would be desirable to know the relation of prohibition of superluminal causation by SEL and by absence of probabilistic counterfactual causal dependence. The difficulty in this problem comes mainly from the fact that SEL is formulated in terms of models of ARQFT considered as a formal language, whereas counterfactual causal dependence is analyzed in terms of possible worlds. Thus, to investigate their relation one has to either reformulate SEL in terms of counterfactuals, which is done in [38], or define counterfactual dependence, and in particular the possible worlds, in terms of models of ARQFT, which is done in [128]. Both approaches seem to distort in some way the content of the original notions of SEL and counterfactual dependence, respectively. The deviations from the original SEL of the various "counterfactualized" SEL notions, which are acknowledged and considered advantaguous in [38], weaken any claim concerning the equivalence of SEL and absence of counterfactual causal dependence. The definition of counterfactual causal dependence in [128] deviates from a Lewisian analysis in the sense that the counterfactual conditionals are not evaluated in a strict manner within the possible world semantics. Thus it seems that the SEL property and the absence of probabilistic counterfactual dependence are different prohibitions of superluminal causation. Further arguments for this independence can be found in [128].

11.3. Bibliographic notes

The problem of logical independence in quantum logic was raised in [129] and [130], where the Definition 11.6 is taken from. The propositions characterizing logical independence are also taken from [129] and [130]. The Schlieder property first appears in [140], the Schlieder-Roos Theorem was proved by Roos in [138] and it is described also in [72]. The Jauch-Piron property is analyzed in the von Neumann algebra framework in [4], [59], [34]; for a discussion of the Jauch-Piron property in more general context see [33]. Lewis' theory of counterfactual probabilistic (in)dependence is applied to the Bell system in [36] and [37]. The Definition 11.10 and all the propositions on the absence of counterfactual probabilistic dependence in ARQFT are taken from [131]. Concerning the different concepts of statistical independence (in the algebraic approach to quantum mechanics) see the comprehensive review [151].

CHAPTER 12

Reichenbach's common cause principle and quantum field theory

As we have seen in Chapter 10, Bell's inequalities are violated in relativistic quantum field theory, and this has the consequence that quantum field theory predicts superluminal correlations i.e. correlations between events as represented by projections lying in local von Neumann algebras belonging to spacelike separated spacetime regions. We also have seen in the previous chapter that an explanation of the superluminal correlations by assuming that there is a direct causal influence between the correlated events is unlikely to be correct, since the correlated algebras do satisfy a number of independence conditions. The aim of this chapter is to investigate another possible way of explaining the correlations in question: the explanation by a probabilistic common cause. Now it is far from obvious what is meant by a "probabilistic common cause" of a probabilistic correlation. We shall utilize here the classic analysis given by Reichenbach in 1956 [136]. Our aim is to specify Reichenbach's notion of a probabilistic common cause in terms of ARQFT in order to raise the problem of whether the superluminal correlations predicted by ARQFT can be causally explained in field theory in the sense of Reichenbach's probabilistic theory of common cause, suitably adapted to ARQFT. In what follows, first we summarize briefly Reichenbach's common cause principle, and in particular his notion of "screening off" (Section 12.1). We shall distinguish two types of screening off: the strong one, in which the causing event actually implies both of the correlated events; and the genuinely probabilistic case, in which the probabilistic cause does not entail any of the correlated events. This is followed in Section 12.2 by an explicit definition of Reichenbach's principle of common cause in ARQFT ("Screening off Principle", Definition 12.1). We wish to stress that we are not able to give an answer to the apparently difficult question (Problem 12.2 in Section 12.2) of whether the superluminal correlations predicted by ARQFT have a probabilistic common cause in general. It is shown in Section 12.2, however, that if each single superluminal correlation predicted by the *vacuum state* between events in $\mathcal{N}(V_1)$ and $\mathcal{N}(V_2)$ has a *genuinely probabilistic* common cause, then the local algebras $\mathcal{N}(V_1)$ and $\mathcal{N}(V_2)$ must be statistically independent in the sense of C^*-independence. It follows then that the existence of truly probabilistic common causes entails that the algebras satisfy a number of equivalent independence conditions (see the concluding remarks).

12.1. Reichenbach's common cause principle

Let A and B be two events and $p(A)$ and $p(B)$ be their probabilities. If the joint probability $p(AB)$ of A and B is greater than the product of the single probabilities, i.e. if

$$p(AB) > p(A)p(B) \tag{12.1}$$

then the events A and B are said to be *correlated*. According to Reichenbach ([136], Section 19), a probabilistic common cause type explanation of a correlation like (12.1) means finding a third event C (cause) such that the following (independent) conditions hold:

$$p(AB|C) = p(A|C)p(B|C) \tag{12.2}$$
$$p(AB|C^\perp) = p(A|C^\perp)p(B|C^\perp) \tag{12.3}$$
$$p(A|C) > p(A|C^\perp) \tag{12.4}$$
$$p(B|C) > p(B|C^\perp) \tag{12.5}$$

where $p(X|Y)$ denotes here the conditional probability of X on condition Y, and it is assumed that none of the probabilities $p(X)$, $(X = A, B, C)$ is equal to zero.

Proposition 12.1 *If the conditional probabilities $p(A|C)$ are defined in the standard way as $p(A|C) = p(AC)/p(C)$ etc., then the conditions (12.2)-(12.5) imply (12.1).*

Proof: One can write

$$p(A) = p(A|C)p(C) + p(A|C^\perp)p(C^\perp) \tag{12.6}$$
$$p(B) = p(B|C)p(C) + p(B|C^\perp)p(C^\perp) \tag{12.7}$$
$$p(AB) = p(AB|C)p(C) + p(AB|C^\perp)p(C^\perp) \tag{12.8}$$

So using (12.2)-(12.3) we have

$$\begin{aligned}
p(AB) - p(A)p(B) &= \\
p(A|C)p(B|C)p(C) &+ \\
p(A|C^\perp)p(B|C^\perp)p(C^\perp) &- \\
[p(A|C)p(C) + p(A|C^\perp)p(C^\perp)] &\times \\
\times [p(B|C)p(C) + p(B|C^\perp)p(C^\perp)] & \\
= p(C)p(C^\perp)[p(A|C) - p(A|C^\perp)] &\times [p(B|C) - p(B|C^\perp)]
\end{aligned}$$

Since both $p(C)$ and $p(C^\perp)$ are assumed to be non-zero, the right hand side of (12.9) is non-zero by (12.4)-(12.5) and the statement follows. □

The condition (12.2) has become known as "screening off", it expresses "... the fact that relative to the cause C the events A and B are mutually independent" ([136] p. 159); that is to say, the common cause event C "screens off" the correlation in the sense that conditionalizing the probability measure p by C, the conditioned probability $p(\bullet|C)$ renders the two events A and B statistically independent. One way to interpret the screening off condition (12.2) is to re-write it as

$$p(A|BC) = p(A|C) \qquad (12.9)$$
$$p(B|AC) = p(B|C) \qquad (12.10)$$

Conditions (12.9)-(12.10) can be read as saying that "knowing the cause C already yields enough information to predict the probability of the event $A(B)$, information on $B(A)$ is redundant".

Notice that there exist two opposite ways the screening off condition (12.2) can be satisfied:

(i) It can happen that, in addition to being a probabilistic common cause, the event C (thought of as an element in a Boolean algebra) is contained both in A and in B, $C \subseteq A$, $C \subseteq B$, and, as another extreme,

(ii) it can also happen that C is a probabilistic cause that is contained neither in A nor in B.

Case (i) means that the event C is not simply a *probabilistic* common cause but a cause that necessarily entails the events A and B, and the screening off condition (12.2) holds then in a trivial way (i.e. with the conditional probabilities all being equal to 1). Given a correlation between A and B, if a probabilistic common cause C can be found such that (in addition to the conditions (12.2)-(12.5)) $C \subseteq A$ and $C \subseteq B$ also is the case, then we say that the correlation can be screened off in the *strong sense*. We refer to the situation (ii) by calling C a *truly (genuinely) probabilistic* common cause.

Note that, given a statistically correlated pair of events A, B in a classical probability space (Ω, μ) with a Boolean algebra Ω of events and a probability measure μ on Ω, a common cause $C(\neq A, B)$ in the sense of Reichenbach's definition does not necessarily exist: The set of events might be too small to contain a common cause. This leads to the question of whether the probability space can be enlarged so that the larger space contains a common cause of a given correlation. More precisely, we have the following problem, which to our best knowledge has not been investigated in the literature, and seems to be open.

Problem 12.1 Let (Ω, μ) be a classical probability space such that for some $A, B \in \Omega$ we have
$$\mu(AB) > \mu(A)\mu(B)$$
Does there exist a probability space (Ω', μ') and an embedding of Ω into Ω' by a Boolean algebra homomorphism h such that
$$\mu(A) = \mu'(h(A)) \qquad A \in \Omega$$
and such that (Ω', μ') contains a common cause $C \neq h(A), h(B)$ of the correlation
$$\begin{aligned} \mu(AB) &= \mu'(h(A)h(B)) \\ > \mu(A)\mu(B) &= \mu'(h(A))\mu'(h(B)) \end{aligned}$$
in the sense of Reichenbach's definition of common cause?

Should it turn out that there is an affirmative answer to the question in the above problem, this would have the philosophical consequence that it is not possible to prove conclusively that the world is is such that distant probabilistic correlations do not have a probabilistic common cause; for one could always argue then that there exist as yet undiscovered events interpretable as (probabilistc) common causes.

Another question in connection with the Reichenbachian scheme of the common cause is the following. Supose that we are given the real numbers

$$r_{AB}, \qquad r_A, r_B$$
$$r_{AB|C}, \qquad r_{A|C}, r_{B|C}$$
$$r_{AB|C^\perp}, \qquad r_{A|C^\perp}, r_{B|C^\perp}$$

and that these numbers satisfy the Reichenbachian conditions (12.2)-(12.5) with the substitutions indicated by the subscripts attached to the numbers. Does there exist then a probability space (Ω, μ) such that there exists events $A, B, C, C^\perp \in \Omega$ with the property

$$\begin{aligned} \mu(AB) &= r_{AB} \\ \mu(A) &= r_A \\ \mu(B) &= r_B \\ \mu(AB|C) &= r_{AB|C} \\ \mu(A|C) &= r_{A|C} \\ \mu(B|C) &= r_{B|C} \\ \mu(AB|C^\perp) &= r_{AB|C^\perp} \\ \mu(A|C^\perp) &= r_{A|C^\perp} \\ \mu(B|C^\perp) &= r_{B|C^\perp} \end{aligned}$$

In other words, the problem is whether any set of numbers satisfying the Reichenbachian conditions can be considered as probabilities in the sense of classical probability theory. Clearly, it is not obvious that the answer is positive, for there are restrictions coming from the assumption that the numbers are probabilities: One has to show that there exist numbers $\mu(C)$, $\mu(C^\perp)$, $\mu(A \wedge C)$ and $\mu(A \wedge C^\perp)$ etc. for all possible combinations of events using the connectives $\wedge$, $\vee$ and the operation $\perp$, and one has to show, further, that these numbers exist consistently, i.e. that μ has the properties of a probability measure.

12.2. Do superluminal correlations have a probabilistic common cause?

As we have seen in Section 10.4, ARQFT predicts superluminal correlations, i.e. correlations of the following sort:

$$\phi(AB) > \phi(A)\phi(B) \tag{12.11}$$

where ϕ is a state on the quasilocal von Neumann algebra $\mathcal{M}$ determined by a covariant net $\{\mathcal{N}(V)\}, V \subseteq M)$ of local observable algebras, and where $A \in \mathcal{N}(V_1)$ and $B \in \mathcal{N}(V_2)$ are projections from the local algebras $\mathcal{N}(V_1)$ and $\mathcal{N}(V_1)$ belonging to spacelike separated spacetime regions V_1 and V_2.

We wish to raise here the problem whether the correlations of the type (12.11) can be explained by finding a common cause in Reichenbach's sense. To make this problem precise, we have to adopt Reichenbach's notion of common cause to the situation in ARQFT. This is done in the next definition.

Definition 12.1 Let V_1 and V_2 be two spacelike separated (open, bounded) spacetime regions, $BLC(V_1)$ and $BLC(V_2)$ be their backward light cones, and $\{\mathcal{N}(V)\}$ be a net of local von Neumann algebras satisfying the standard axioms. We say that the pair of algebras $\mathcal{N}(V_1), \mathcal{N}(V_2)$ *satisfies (Reichenbach's) Screening off Principle* if and only if for any state ϕ over the quasilocal algebra $\mathcal{N}$ and for any pair of projections $A \in \mathcal{N}(V_1)$ $B \in \mathcal{N}(V_2)$ we have the following: if

$$\phi(AB) > \phi(A)\phi(B)$$

then there exists a projection C in the von Neumann algebra $\mathcal{N}(V)$ that is associated with a region V lying within the intersection

$$BLC(V_1) \cap BLC(V_2)$$

such that

$$\phi(C) \neq 0 \neq \phi(C^\perp)$$

and C satisfies the following conditions:

(i) C commutes with both A and B
(ii) the conditions below (analogous to (12.2), (12.3), (12.4) and (12.5)) hold:

$$\frac{\phi(ABC)}{\phi(C)} = \frac{\phi(AC)}{\phi(C)}\frac{\phi(BC)}{\phi(C)} \qquad (12.12)$$

$$\frac{\phi(ABC^\perp)}{\phi(C^\perp)} = \frac{\phi(AC^\perp)}{\phi(C^\perp)}\frac{\phi(BC^\perp)}{\phi(C^\perp)} \qquad (12.13)$$

$$\frac{\phi(AC)}{\phi(C)} > \frac{\phi(AC^\perp)}{\phi(C^\perp)} \qquad (12.14)$$

$$\frac{\phi(BC)}{\phi(C)} > \frac{\phi(BC^\perp)}{\phi(C^\perp)} \qquad (12.15)$$

We say that the *Screening off Principle holds* in ARQFT iff for every pair of spacelike separated spacetime regions V_1, V_2 the Screening off Principle holds for the pair $\mathcal{N}(V_1), \mathcal{N}(V_2)$. Just like in the case of Reichenbach's formulation, one can distinguish the strong and genuinely probabilistic versions of probabilistic common cause in ARQFT, and one can speak accordingly of the Screening off Principle holding in ARQFT in the strong and genuinely probabilistic sense.

The Screening off Principle as specified above differs slightly from Reichenbach's in two respects: First, since ARQFT is a non-commutative theory, one has to require explicitly the commutativity of the events involved – unless one is willing to expand Reichenbach's scheme and replace it by a theory of "non-commutative screening off", involving non-commutative conditionalization, which we do not wish to consider here. (See the paper [155] for an analysis of some technical difficulties concerning the generalization of Reichenbach's scheme to non-distributive event structures.) Second, the common cause event C is required in the above definition to lie in the common causal past of the two correlated events. This latter condition was not part of Reichenbach's original theory. It could not be because that theory was not formulated within the framework of Minkowski spacetime. But it is clear that as soon as one is in a theory where there is an underlying causal structure to consider, like in ARQFT, the condition that C be causally not disconnected from either A or B must be required, otherwise one could hardly talk about a *common* cause explanation in *relativistic* sense.

We are now in the position to ask

Problem 12.2 *Does ARQFT satisfy the Screening off Principle?*

As we have indicated already, we are not able to answer this question, nor do we know of any result that would give a partial answer, positive or negative. What will be seen below is that the existence of a genuinely probabilistic common cause of every vacuum correlation entails C^*-independence of the algebras involved.

Proposition 12.2 *Let V_1, V_2 be two open, bounded spacelike separated spacetime regions and $\mathcal{N}(V_1), \mathcal{N}(V_2)$ be the two von Neumann algebras in a net of von Neumann algebras in an irreducible vacuum representation of a net of local C^*-algebras satisfying the standard conditions described in Chapter 10. If each single correlation between projections of $\mathcal{N}(V_1), \mathcal{N}(V_2)$ predicted by the vacuum state has a genuinely probabilistic common cause explanation in the sense described in the Definition 12.1, then the two algebras $\mathcal{N}(V_1), \mathcal{N}(V_2)$ are C^*-independent.*

Proof: The statement is an easy consequence of the powerful, non-trivial Schlieder-Roos and Reeh-Schlieder theorems. Recall that the Schlieder-Roos theorem (Proposition 10.5)) says that if $\mathcal{A}_1$ and $\mathcal{A}_2$ are mutually commuting C^*-algebras (i.e. $XY = YX$ for all $X \in \mathcal{A}_1, Y \in \mathcal{A}_2$), then C^*-independence of $\mathcal{A}_1, \mathcal{A}_2$ is equivalent to the following condition ("Schlieder property"): $XY \neq 0$ whenever $0 \neq X \in \mathcal{A}_1, 0 \neq Y \in \mathcal{A}_2$. The Reeh-Schlieder theorem (Proposition 10.1) says that the vacuum vector Ω_0 is both cyclic and separating for any local algebra belonging to a region V with non-empty causal complement; in other words, no non-zero positive element in $\mathcal{A}(V)$ can annihilate the vacuum vector: if $0 \leq X \in \mathcal{N}(V)$ and $X\Omega_0 = 0$ then $X = 0$. By the Schlieder-Roos theorem it is enough to show that the assumptions in the proposition imply the Schlieder property. The reasoning preceding Proposition 11.5 shows that to prove the Schlieder property, it is enough to prove it for projections only; so let $A \in \mathcal{A}(V_1)$ and $B \in \mathcal{A}(V_2)$ be arbitrary non-zero projections. We must show that $AB \neq 0$. Consider now the vacuum state

$$X \mapsto \langle \Omega_0, X\Omega_0 \rangle = \phi_0(X)$$

One of the following three equations holds.

$$\langle \Omega_0, AB\Omega_0 \rangle > \langle \Omega_0, A\Omega_0 \rangle \langle \Omega_0, B\Omega_0 \rangle \quad (12.16)$$
$$\langle \Omega_0, AB\Omega_0 \rangle = \langle \Omega_0, A\Omega_0 \rangle \langle \Omega_0, B\Omega_0 \rangle \quad (12.17)$$
$$\langle \Omega_0, AB\Omega_0 \rangle < \langle \Omega_0, A\Omega_0 \rangle \langle \Omega_0, B\Omega_0 \rangle \quad (12.18)$$

The right hand sides of all of the above equations is strictly positive by the Reeh-Schlieder theorem, therefore if either (12.16) or (12.17) is the case then

$$\langle \Omega_0, AB\Omega_0 \rangle > 0$$

and so $AB \neq 0$. If equation (12.18) is the case, then one checks easily that

$$\langle \Omega_0, A^\perp B\Omega_0 \rangle > \langle \Omega_0, A^\perp \Omega_0 \rangle \langle \Omega_0, B\Omega_0 \rangle \qquad (12.19)$$

By assumption there is a genuinely probabilistic common cause of the correlation (12.19), i.e. there exists a C projection in a local algebra $\mathcal{A}(V)$, where

$$V \subseteq BLC(V_1) \cap BLC(V_2)$$

such that C commutes with both A and B, satisfying $C \not\subseteq A^\perp$ and $C \not\subseteq B$, and such that

$$\frac{\langle \Omega_0, A^\perp BC\Omega_0 \rangle}{\langle \Omega_0, C\Omega_0 \rangle} = \frac{\langle \Omega_0, A^\perp C\Omega_0 \rangle}{\langle \Omega_0, C\Omega_0 \rangle} \frac{\langle \Omega_0, BC\Omega_0 \rangle}{\langle \Omega_0, C\Omega_0 \rangle} \qquad (12.20)$$

$$\frac{\langle \Omega_0, A^\perp BC^\perp \Omega_0 \rangle}{\langle \Omega_0, C^\perp \Omega_0 \rangle} = \frac{\langle \Omega_0, A^\perp C^\perp \Omega_0 \rangle}{\langle \Omega_0, C^\perp \Omega_0 \rangle} \frac{\langle \Omega_0, BC^\perp \Omega_0 \rangle}{\langle \Omega_0, C^\perp \Omega_0 \rangle} \qquad (12.21)$$

$$\frac{\langle \Omega_0, A^\perp C\Omega_0 \rangle}{\langle \Omega_0, C\Omega_0 \rangle} > \frac{\langle \Omega_0, A^\perp C^\perp \Omega_0 \rangle}{\langle \Omega_0, C\Omega_0 \rangle} \qquad (12.22)$$

$$\frac{\langle \Omega_0, BC\Omega_0 \rangle}{\langle \Omega_0, C\Omega_0 \rangle} > \frac{\langle \Omega_0, BC^\perp \Omega_0 \rangle}{\langle \Omega_0, C\Omega_0 \rangle} \qquad (12.23)$$

By an elementary rewriting of (12.20) one can verify easily that the following also holds:

$$\frac{\langle \Omega_0, ABC\Omega_0 \rangle}{\langle \Omega_0, C\Omega_0 \rangle} = \frac{\langle \Omega_0, AC\Omega_0 \rangle}{\langle \Omega_0, C\Omega_0 \rangle} \frac{\langle \Omega_0, BC\Omega_0 \rangle}{\langle \Omega_0, C\Omega_0 \rangle} \qquad (12.24)$$

$\langle \Omega_0, BC\Omega_0 \rangle$ is non-zero by (12.23), hence, if $\langle \Omega_0, AC\Omega_0 \rangle$ is shown to be non-zero, then the right hand side of (12.24) is not equal to zero, and the proof is then complete. If $\langle \Omega_0, AC\Omega_0 \rangle$ were equal to zero, then (since AC is a projector, hence non-negative) $AC = 0$ would follow by the Reeh-Schlieder Theorem; but $AC = 0$ implies $C \subseteq A^\perp$, which can not be the case, since C was assumed to be a genuinely probabilistic common cause of the correlation (12.19). □

Statistical independence is a property that is typically expected to hold for local algebras pertaining to spacelike separated, i.e. causally disconnected spacetime regions. The Screening off Principle, on the other hand, involves causally connected regions and algebras. The Proposition 12.2 connects the two notions, and it shows that C^*-independence of spacelike separated algebras is necessary for the Screening off Principle (in the genuinely probabilistic sense) to hold in ARQFT. (It is also clear from the proof that the above proposition remains valid by replacing the vacuum state by any other *faithful* state.) Since in the context of ARQFT

C^*-independence, W^*-independence and strict locality are equivalent (Proposition 10.6), it follows then that validity of the Screening off Principle in ARQFT (in the probabilistic sense) implies both W^*-independence and strict locality of the local algebras confined in spacelike separated spacetime regions.

However, the proof of the Proposition 12.2 also indicates that C^*-independence (hence also W^*-independence and strict locality) is unlikely to be sufficient for the Screening off Principle to hold: One of the properties of the probabilistic common cause, namely that the common cause C belongs to the common causal past of the correlated events, was not used in inferring the C^*-independence property.

Since the Screening off Principle appears to be stronger than C^*-independence, a natural question is whether it implies that stronger independence conditions hold. Since it is known that W^*-independence in the product sense (hence also the so-called "split property", see Definitions 10.5 and 10.7 and the Proposition 10.8) is a strictly stronger independence condition than W^*-independence, if W^*-independence in the product sense or the split property could be inferred from the Screening off Principle, then one could conclude that the Screening off Principle does not hold in general, since it is known that the split property fails for *tangent* spacetime regions. It is not known, whether the Screening off Principle implies any of the stronger statistical independence conditions. Most pressing would be to know, however, whether the Screening off Principle can hold at all, at least for some pairs of spacelike separated spacetime regions. It is not inconceivable that the Screening off Principle is independent of the other standard axioms formulated on the net of local von Neumann algebras. This would mean that those axioms are not rich enough to characterize exhaustively the causal structure of the local algebras, for they leave it open whether the world of quantum fields is such that distant correlations have a common cause or not.

The results on the violation of Bell's inequality in ARQFT imply that, if for a given state ϕ there exists a *single, common* probabilistic common cause C (in the sense of Definition) of *all* correlations predicted by ϕ, then the C-conditioned state $\phi(\bullet|C)$ is a product state across the algebras $\mathcal{A}(V_1), \mathcal{A}(V_2)$. Since a product state satisfies Bell's inequality, and since for tangent spacelike separated wedge and double cone regions *every* normal state maximally violates Bell's inequality, there exists no normal state over local algebras in the said regions such that the correlations predicted by it have a *common* common cause. But the assumption that *all* the superluminal correlations predicted by a given state in ARQFT have a *common* common cause, seems totally unwarranted. Not only isn't there anything in the Reichenbachian notion of common cause that would justify this assumption, the common cause principle doesn't even seem to contain any hint as to how the different common causes

$$C^{A',B'}, C^{A'',B''} \ldots$$

of different correlated pairs

$$(A',B');(A'',B'') \ldots$$

(possibly containing even incompatible elements) might be related to each other. This dependence of the common cause on the pair of the correlated events and the unrelatedness of the causes of correlations of different event-pairs not simply blocks the inference from the assumption of existence of common causes to the value of the Bell correlation, but it makes unclear in which state one should check the value of the Bell correlation: given a state, the vacuum ϕ_0 say, and assuming that there exist probabilistic common causes

$$C^{A',B'}, C^{A'',B''} \ldots$$

of all the correlated pairs

$$(A',B');(A'',B'') \ldots$$

we have the conditioned states

$$\phi_0(\bullet|C^{A',B'}), \phi_0(\bullet|C^{A'',B''}) \ldots$$

Which of these states should satisfy Bell's inequality (10.9)? In fact we know (since all these states are normal) that *each* violates Bell's inequality (10.9) (for complementary wedges and spacelike separated tangent double cones). But why shouldn't they – assuming only (12.12)-(12.15) to hold with $A', B', C'; A'', B'', C'' \ldots$?

In short, under the present specification of Bell's inequality and of the concept of Reichenbachian common cause, it is impossible to give meaning to the claim "Bell's inequality is implied by Reichenbach's common cause principle"; hence, on the present interpretation, violation of Bell's inequality does not imply the impossibility of Reichenbachian common causes of superluminal correlations. Whether such (not common) Reichenbachian probabilistic common causes exist in ARQFT remains an open question.

12.3. Bibliographic notes

Reichenbach's notion of probabilistic common cause was formulated in [136], and it was formulated without regard to quantum correlations. For a philosophical critique of Reichenbach's notion of probabilistic common cause see [139]. The Definition 12.1 and Proposition 12.2 are taken from [133]. The relation of the screening off property to Bell's inequality (in non-relativistic quantum mechanics) is analyzed by Butterfield in [35] and by van Fraassen in [162]. The conclusion in both of these papers is that the assumption of existence of Reichenbachian common causes of quantum correlations *does* imply Bell's inequality. In both papers common cause is understood, however, as *common* common cause. In a recent paper Belnap and Szabó have proved that the non-probabilistic superluminal correlations occurring in the Greenberger-Horne-Zeilinger (GHZ) situation do not have a non-probabilistic *common* common cause [16], where the notion of (non-probabilistic) common cause is formulated in terms of the branching spacetime theory [15]. Remarkably, in that paper it also remains open, however, whether a *non-common* common cause (in the non-probabilistic sense) for the GHZ correlations exists.

References

1. L. Accardi, F. Frigerio and V. Gorini (eds.): *Quantum Probability and Applications to the Quantum Theory of Irreversible Processes*, Lecture Notes in Mathematics, Vol. 1055, Springer Verlag, Berlin-Heidelberg, 1984
2. L. Accardi and W. von Waldenfells, (eds.): *Quantum Probability and Applications*, Lecture Notes in Mathematics, Vol. 1303, Springer Verlag, Berlin-Heidelberg, 1988
3. L. Accardi and W. von Weidenfells (eds.): *Quantum Probability and Applications*, Lecture Notes in Mathematics, Vol. 1442, Springer Verlag, Berlin-Heidelberg, 1990
4. A. Amann: Jauch-Piron states in W^*-algebraic quantum mechanics
   Journal of Mathematical Physics **28** (1989) 2384-2389
5. F.C. Andrews: *Equilibrium Statistical Mechanics*, New York, John Wiley, 1975
6. A. Aspect, P. Grangier and G. Roger: Experimental tests of realistic local theoreis via Bell's theorem
   Physical Review Letters **47** (1981) 460-467
7. A. Aspect, P. Grangier and G. Roger: Experimental realization of Einstein-Podolsky-Rosen-Bohm *Gedankenexperiment*: A new violation of Bell's inequalities
   Physical Review Letters **48** (1982) 91-94
8. A. Aspect, J. Dalibard and G. Roger: Experimental tests of Bell's inequalities using time-varying analyzers
   Physical Review Letters **49** (1982) 1804-1807
9. F.J. Belinfante: *A Survey of Hidden Variable Theories*, Pergamon Press, Oxford, 1973
10. J.S. Bell (1964): On the Einstein-Podolsky-Rosen paradox
    Physics **1** (1964) 196-200
    (reprinted in [13])
11. J.S. Bell (1966): On the problem of hidden variables in quantum mechanics
    Reviews in Modern Physics **38** (1966) 447-475
    (reprinted in [13])
12. J.S. Bell: Introduction to the hidden variable question
    in *Proceedings of the Interantional School of Physics "Enrico Fermi", Course 49, "Foundations of Quantum Mechanics"*, B. Espagnat (ed.), Academic Press, New York-London, 1971, 171-181
    (reprinted in [13])
13. J.S. Bell: *Speakable and unspeakable in quantum mechanics*, Cambridge University Press, Cambridge, 1987
14. J.L. Bell and R.K. Clifton: QuasiBoolean algebras and simultaneously definite properties in quantum mechanics
    International Journal of Theoretical Physics **34** (1985) 2409-2421
15. N. Belnap: Branching space-time
    Synthese **92** (1992) 385-434
16. N. Belnap and L.E. Szabó: Branching space-time analysis of the GHZ theorem
    Foundations of Physics **26** (1996) 989-1002
17. E. Beltrametti and Bas C. Fraassen (eds.): *Current Issues in Quantum Logic*, Plenum Press, New York, 1981
18. E.G. Beltrametti and G. Cassinelli: *The logic of quantum mechanics*, Addison

Wesley, Massachusetts, 1981
19. G. Birkhoff: *Lattice Theory*, 3rd ed., Amer. Math. Soc. Colloq. Publ. Vol. 25 (1966)
20. G. Birkhoff: Lattices in applied mathematics
    in [46] 155-184
21. G. Birkhoff and J. von Neumann: The logic of quantum mechanics
    Annals of Mathematics **37** (1936) 823-843
    in [178] 105-125
22. D. Bohm: A suggested interpretation of the quantum theory in terms of hidden variables I,II
    Physical Review **85** (1952) 166-179, 180-193
23. D. Bohm and J. Bub: A proposed solution of the measurement problem in quantum mechanics by a hidden variable theory
    Reviews in Modern Physics **38** (1966) 453-469
24. D. Bohm and J. Bub: A refutation of the proof by Jauch and Piron that hidden variables can be excluded in quantum mechanics
    Reviews in Modern Physics **38** (1966) 470-475
25. D. Bohm and J. Bub: On hidden variables - A reply to comments by Jauch and Piron and by Gudder
    Reviews in Modern Physics **40** (1968) 235-236
26. M. Born: Quantenmechanik der Stossvorgänge
    Zeitschrift für Physik **38** (1926) 803-827
27. O. Bratteli and D.W. Robinson: *Operator Algebras and Quantum Statistical Mechanics. Vol.I. C^*-and $W*$-algebras, Symmetry Groups, Decomposition of States*, Springer, New York-Berlin-Heidelberg, 1979
28. O. Bratteli and D.W. Robinson: *Operator Algebras and Quantum Statistical Mechanics. Vol.II. Equilibrium States, Models in Quantum Statistical Mechanics*, Springer, New York-Berlin-Heidelberg, 1981
29. F. Brody and T. Vámos (eds.): *The Neumann Compendium*, World Scientific Series of 20th Century Mathematics Vol. I., World Scientific, Singapore, 1995
30. J. Bub: *The interpretation of quantum mechanics*, Reidel, Dordrecht, 1974
31. J. Bub: What does quantum logic explain?
    in [17] 89-100
32. J. Bub: Hidden variables and quantum mechanics – a sceptical review
    Erkenntnis **16** (1981) 275-293
33. L.J. Bunce, M. Navara, P. Pták and J.D.M Wright: Quantum logics with Jauch-Piron states
    Quarterly Journal of Mathematics Oxford (2) **36** (1985) 261-271
34. L.J. Bunce and J. Hamhalter: Jauch-Piron states on von Neumann algebras
    Mathematische Zeitschrift **215** (1994) 491-502
35. J. Butterfield: A space-time approach to the Bell inequality
    in [43] 114-144
36. J. Butterfield: David Lewis Meets John Bell
    Philosophy of Science **59** (1992) 26-43
37. J. Butterfield: Bell's Theorem: What it Takes
    British Journal for the Philosophy of Science **58** (1992) 41-83
38. J. Butterfield: Outcome dependence and stochastic Einstein nonlocality
    in [107] 385-424
39. J. Butterfield: Vacuum correlations and outcome dependence in algebraic quantum field theory
    in [53] 768-785
40. J.F. Clauser, M.A. Horne and A. Shimony: Bell's theorem: experimantal tests and implications
    Reports on Progress in Physics **41** (1978) 1881-1927
41. R. Clifton: Getting contextual and nonlocal elements-of-reality the easy way
    American Journal of Physics **61** (1993) 443-447

REFERENCES

42. R. Clifton (ed.): *Perspectives on Quantum Reality: Relativistic, Non-Relativistic and Field Theoretic*, Kluwer Academic Publishers, 1966
43. J. Cushing and E. McMullin (eds.): *Philosophical Consequences of Quantum Theory*, University of Notre Dame Press, Notre Dame, IN, 1989
44. E.B. Davies: *Quantum Theory of Open Systems*, Academic Press, 1976
45. D. Deutsch: Uncertainty in quantum measurements
    Physical Review Letters **50** (1983) 631-633
46. R.P. Dilworth (ed.): *Lattice Theory* (Proceedings of the Second Symposium in Pure Mathematics of the American Mathematical Society, April 1959), American Mathematical Society, Providence, 1961
47. W. Driessler: On the type of local algebras in quantum field theory
    Communications in Mathematical Physics **53** (1977) 295-297
48. W. Driessler, S.J. Summers and E.H. Wichman: On the connection between quantum fields and von Neumann algebras of local operators
    Communications in Mathematical Physics **105** (1986) 49-84
49. G. Emch: *Algebraic Methods in Statistical Physics and Quantum Field Theory*, Wiley Interscience, New York, 1972
50. G. Fleming and J. Butterfield: Is there superluminal causation in quantum theory?
    in [99] 203-207
51. M. Florig and S.J. Summers: On the statistical independence of algebras of observables
    Journal of Mathematical Physics **3** (1997) 1318-1328
52. A.M. Gleason: Measures on the closed subspaces of a Hilbert space
    Journal of Mathematics and Mechanics **6** (1957) 885-893
    in [70] 123-133
53. D.M. Greenberger and A. Zeilinger (eds.): *Fundamental Problems in Quantum Theory*, Annals of the New York Academy of Sciences, **755** (1994)
54. S.P. Gudder: Hidden variables in quantum mechanics reconsidered
    Reviews in Modern Physics **40** (1968) 229-231
55. S.P. Gudder: On hidden variable theories
    Journal of Mathematical Physics **11** (1970) 431-436
56. R. Haag: *Local Quantum Physics. Fields, Particles, Algebras*, Springer Verlag, Berlin, 1992
57. R. Haag and D. Kastler: An algebraic approach to quantum field theory
    Journal of Mathematical Physics **5** (1964) 848-861
58. P.R. Halmos: *Introduction to Hilbert Space and the Theory of Spectral Multiplicity*, Chelsa, New York, 1957
59. J. Hamhalter: Pure Jauch-Piron states on von Neumann algebras
    Annales de l'Institut Henri Poincaré **58** (1993) 173-187
60. G. Hardegree: The conditional in quantum logic
    in [154] 55-72
61. G. Hardegree: The conditional in abstract and concrete quantum logic
    in [71] 49-108
62. G. Hardegree: Material implication in orthomodular lattices
    Notre Dame Journal of Formal Logic **22** (1981) 163-182
63. G. Hardy, J. Littlewood and G. Pólya: *Inequalities*, Cambridge University Press, Cambridge, 1934
64. W.L. Harper and C.A. Hooker (eds.): *Foundations of Probability Theory, Statistical Inference and Statistical Theories of Science*, D. Reidel Publishing Co. Dordrecht, Holland, 1976
65. A. Hartkämper and H. Neumann (eds.): *Foundations of Quantum Mechanics and Ordered Linear Spaces*, Lecture Notes in Physics, Vol. 29., Springer, Heidelberg, 1974
66. G. Hellman: Einstein and Bell: Strengthening the case for microphysical randomness

Synthese **53** (1982) 445-460
67. G. Hellman: Stochastic Einstein-locality and the Bell theorems
Synthese **53** (1982) 461-504.
68. D. Hilbert, L. Nordheim and J. von Neumann: Über die Grundlagen der Quantenmechanik
Mathematische Annalen **98** (1927) 1-30
in [175] 104-133
69. S.S. Holland Jr.: The current interest in orthomoduar lattices
in [70] 437-496
70. C.A. Hooker (ed.): *The Logico-Algebraic Approach to Quantum Mechanics. Vol. I. Historical Evolution*, D. Reidel Publishing Co. Dordrecht Holland, 1975
71. C.A. Hooker (ed.): *The Logico-Algebraic Approach to Quantum Mechanics. Vol. II. Contemporary Consolidation*, D. Reidel Publishing Co. Dordrecht Holland, 1975
72. S. Horuzhy: *Introduction to Algebraic Quantum Field Theory*, Kluwer, 1990
73. M. Jammer: *The Philosophy of Quantum Mechanics*, Wiley Interscience, New York, 1974
74. J.M. Jauch and C. Piron: Can hidden variables be excluded in quantum mechanics?
Helvetica Physica Acta **36** (1963) 827-837
75. J. M. Jauch: *Foundations of Quantum Mechanics*, Addison Wesley, 1968
76. J.M. Jauch and C. Piron: Hidden variables revisited
Reviews in Modern Physics **40** (1968) 228-229
77. E.T. Jaynes: The well-posed problem
Foundations of Physics **3** (1973) 477-493
78. V.F.R. Jones: Index for subfactors
Inventiones Mathematicae **72** (1983) 1-25
79. P. Jordan: Über Verallgemeinerungsmöglichkeiten des Formalismus der Quantenmechanik
Göttinger Nachrichten (1933) 209
80. P. Jordan: Über die Multiplikation quantenmechanischen Grössen
Zeitschrift für Physik **80** (1933) 285-291
81. P. Jordan, E. Wigner and J. von Neumann: On an algebraic generalization of the quantum mechanical formalism
Annals of Mathematics **35** (1934) 29-64
in [176] 409-444
82. Richard V. Kadison and John R. Ringrose: *Fundamentals of the Theory of Operator Algebras. Vol. I. Elementary Theory*, Academic Press Inc., New York, 1983
83. Richard V. Kadison and John R. Ringrose: *Fundamentals of the Theory of Operator Algebras. Vol. II. Advanced Theory*, Academic Press Inc., New York, 1986
84. G. Kalmbach: *Orthomodular Lattices*, Academic Press, London, 1983
85. S. Kochen and E. Specker: The problem of hidden variables in quantum mechanics
Journal of Mathematics and Mechanics **17** (1967) 59-67
in [70] 293-328
86. K. Kraus: General quantum field theories and strict locality
Zeitschrift für Physik **191** (1964) 1-12
87. K. Kraus: Complementary observables and uncertainty relations
Physical Review D **35** (1987) 3070-3075
88. P. Kruszynski: Extensions of Gleason theorem
in [1] 210-227
89. P. Lahti and P. Mittelstaedt (eds.): *Symposium on the Foundations of Modern Physics*, World Scientific, Singapore, 1985
90. P. Lahti, Kari Ylinen: On total noncommutativity in quantum mechanics
Journal of Mathematical Physics **28** (1987) 2614-2617
91. L.J. Landau: On the violation of Bell's inequality in quantum theory
Physics Letters **A120** (1987) 54-56

92. D. Lewis: *Counterfactuals*, Blackwell, Oxford, 1973
93. D. Lewis: *Collected Papers. Volume II.*, Oxford University Press, Oxford, 1986
94. M. Loéve: *Probability Theory II*, 4th edition, Springer, New York, 1978
95. H. Maassen: A discrete entropic uncertainty relation
    in [3] 263-266
96. H. Maassen and J.B.M. Uffink: Generalized entropic uncertainty relations
    Physical Review Letters **60** (1988) 1103-1106
97. F. Maeda and S. Maeda: *Theory of Symmetric Lattices*, Springer Verlag, Berlin, Heidelberg, New York, 1970
98. T. Matolcsi and S. Székely: *Matematikai Fizika. Egyetemi jegyzet.*, Tankönyvkiadó, Budapest, 1977
99. A. Van Der Merwe, F. Selleri and G. Tarozzi (eds.): *Bell's Theorem and the Foundations of Modern Physics*, World Scientific, Singapore, 1992
100. Richard von Mises: *Probability, Statistics and Truth* (second English edition of *Wahrscheinlichkeit, Statistik und Wahrheit*, Springer, 1928), Dover Publications, New York, 1981
101. B. Misra: When can hidden variables be excluded in quantum mechanics?
     Nuovo Cimento **47A** (1967) 841-859
102. P. Mittelstaedt: On the interpretation of the lattice of subspaces of Hilbert space as a propositional calculus
     Zeitschrift für Naturforschung **27a** (1972) 1358-1362
103. F. Muller and J. Butterfield: Is algebraic relativistic quantum field theory stochastic Einstein local?
     Philosophy of Science **61** (1994) 457-474
104. F. J. Murray and J. von Neumann: On rings of operators
     Annals of Mathematics **37** (1936) 116-229
     in [177] 6-119
105. W. Ochs: Some comments on the concept of state in quantum mechanics
     Erkenntnis **16** (1981) 339-356
106. M.H. Partovi: Entropic formulation of uncertainty in quantum measurements
     Physical Review Letters **50** (1983) 1883-1885
107. D. Parwitz and D. Westerdahl (eds.): *Logic and Philosophy of Science in Uppsala*, Kluwer, Dordrecht, Holland, 1994
108. M. Pavicic: Bibliography on quantum logic
     International Journal of Theoretical Physics **31** (1992) 373-461
109. D. Petz: Conditional expectation in quantum probability
     in [2] 251-260
110. D. Petz and M. Ohya: *Quantum Entropy and its Use*, (Texts and Monographs in Physics), Springer, 1993
111. D. Petz and M. Rédei: Legacy of von Neumann in the theory of operator algebras
     in [29] 163-181
112. C. Piron: *Foundations of Quantum Physics*, Benjamin, Reading, Massachussetts, 1976
113. I. Pitowsky: *Quantum Probability – Quantum Logic*, Lecture Notes in Physics, Vol. 321, Springer Verlag, Berlin-Heidelberg-New York, 1990
114. R. Powers: Existence of uncountable number of non isomorphic type III factors
     Annals of Mathematics **86** (1967) 138-171
115. E. Prugovecki: *Quantum Mechanics in Hilbert Space*, Academic Press, New York, 1971
116. P. Pták and S. Pulmannová: *Orthomodular Structures as Quantum Logic*, Kluwer Academic Publishers, Dordrecht, Boston, London, 1991
117. H. Putnam: Is logic empirical?
     in [71] 181-206
118. M. Rédei: Note on an argument of W. Ochs against the ignorance interpretation of state in quantum mechanics

Erkenntnis **23** (1985) 143-148
119. M. Rédei: Conditions excluding the existence of approximate hidden variables
Physics Letters **A110** (1985) 15-16
120. M. Rédei: Nonexistence of hidden variables in the algebraic approach
Foundations of Physics **16** (1986) 807-815
121. M. Rédei: On the problem of local hidden variables in algebraic quantum mechanics
Journal of Mathematical Physics **28** (1987) 833-835
122. M. Rédei: Reformulation of the hidden variable problem using entropic measure of uncertainty
Synthese **73** (1987) 371-379
123. M. Rédei: The hidden variable problem in algebraic relativistic quantum field theory
Journal of Mathematical Physics **30** (1989) 461-463
124. M. Rédei: Stochastic irreducibility of spin
Physics Letters **A134** (1989) 354-356
125. M. Rédei: Quantum conditional probabilities are not probabilities of quantum conditional
Physics Letters **A139** (1989) 287-290
126. M. Rédei: Bell's inequalities, relativistic quantum field theory and the problem of hidden variables
Philosophy of Science **58** (1991) 628-638
127. M. Rédei: When can non-commutative statistical inference be Bayesian?
International Studies in the Philosophy of Science **6** (1992) 129-132
128. M. Rédei: Are prohibitions of superluminal causation by stochastic Einstein locality and by absence of Lewisian probabilistic counterfactual causation equivalent?
Philosophy of Science **60** (1993) 608-618
129. M. Rédei: Logical independence in quantum logic
Foundations of Physics **25** (1995) 411-422
130. M. Rédei: Logically independent von Neumann lattices
International Journal of Theoretical Physics **34** (1995) 1711-1718
131. M. Rédei: Is there counterfactual superluminal causation in relativistic quantum field theory?
in [42] 29-42
132. M. Rédei: Why J. von Neumann did not like the Hilbert space formalism of quantum mechanics (and what he liked instead)
Studies in the History and Philosophy of Modern Physics **27** (1996) 493-510
133. M. Rédei: Reichenbach's common cause principle and quantum field theory
Foundations of Physics (forthcoming)
134. M. Redhead: *Incompleteness, Non-locality and Realism: Prolegomenon to the Philosophy of Quantum Mechanics*, Claredon Press, Oxford, 1987
135. W. Rehder: Conditions for probabilities of conditionals to be conditional probabilities
Synthese **53** (1982) 439-443
136. H. Reichenbach: *The Direction of Time*, University of California Press, Los Angeles, 1956
137. M. Riesz: Sur les maxima des formes bilinéaires et sur les fonctionelles linéaires
Acta Mathematica **49** (1927) 465-497
138. H. Roos: Independence of local algebras in quantum field theory,
Communications in Mathematical Physics **16** (1970) 238-246
139. W.C. Salmon: Probabilistic causality
Pacific Philosophical Quarterly **61** (1980) 50-74
140. S. Schlieder: Einige Bemerkungen über Projektionsoperatoren
Communications in Mathematical Physics **13** (1969) 216-225
141. J. Schwinger: Unitary operator bases
Proceedings of the National Academy of Sciences **46** (1960) 570-579

142. E. Störmer: Positive linear maps of operator algebras
Acta Mathematica **110** (1963) 233-278
143. E. Störmer: Positive linear maps of C^*-algebras
in [65] 85-106
144. S. Stratila and L. Zsidó: *Lectures on Von Neumann Algebras*, Abacus Press, Turnbridge Wells, Kent, 1979
145. S.J. Summers and R. Werner: The vacuum violates Bell's inequalities
Physics Letters **A110** (1985) 257-259
146. S.J. Summers and R. Werner: Maximal violation of Bell's inequalities is generic in quantum field theory
Communications in Mathematical Physics **110** (1987) 247-259
147. S.J. Summers and R. Werner: Bell's inequalities and quantum field theory.I. General setting.
Journal of Mathematical Physics **28** (1987) 2440-2447
148. S.J. Summers and R. Werner: Bell's inequalities and quantum field theory.II. Bell's inequalities are maximally violated in the vacuum
Journal of Mathematical Physics **28** (1987) 2448-2456
149. S.J. Summers and R. Werner: Maximal violation of Bell's inequalities for algebras of observables in tangent spacetime regions
Annales de l'Institut Henri Poincaré – Physique theorique **49** (1988) 215-243
150. S.J. Summers: Bell's inequalities and quantum field theory
in [3] 393-413
151. S.J. Summers: On the independence of local algebras in quantum field theory
Reviews in Mathematical Physics **2** (1990) 201-247
152. S.J. Summers and R. Werner: On Bell's inequalities and algebraic invariants
Letters in Mathematical Physics **33** (1995) 321-334
153. V.S. Sunders: *An Invitation to von Neumann Algebras*, Springer Verlag, New York, 1987
154. P. Suppes (ed.): *Logic and Probability in Quantum Mechanics*, D. Reidel Publishing Co. Dordrecht, Holland, 1976
155. G. Szabó: Reichenbach's common cause definition on Hilbert lattices
submitted for publication
156. L.E. Szabó: Is quantum mechanics compatible with a deterministic universe? Two interpretations of quantum probabilities.
Foundations of Physics Letters **8** (1995) 421-440
157. M. Takesaki: Conditional expectations in von Neumann algebras
Journal of Functional Analysis **9** (1972) 306-321
158. M. Takesaki: *Theory of Operator Algebras I.*, Springer Verlag, New York, 1979
159. J.B.M. Uffink: *Measures of Uncertainty and the Uncertainty Principle*, PhD Thesis, Utrecht University, 1990
160. B.C. Van Fraassen: The labyrinth of quantum logics
in *Boston Studies in the Philosophy of Science*, vol. XIII., D.Reidel Publishing Co., Dordrecht, Holland, 1974, 224-254
in [70] 577-607
161. B.C. Van Fraassen: When is a correlation not a mystery?
in [89] 113-128
162. B.C. Van Fraassen: The charybdis of realism: epistemological implications of Bell's inequality
in [43] 97-113
163. B.C. Van Fraassen: *Quantum Mechanics: An Empiricist View*, Claredon Press, Oxford, 1991
164. V. Varadarajan: *Geometry of Quantum Theory I,II*, van Nostrand, Princeton, 1968, 1970
165. J. von Neumann: Mathematische Begründung der Quantenmechanik
Göttinger Nachrichten **1** (1927) 1-57

in [175] 151-207
166. J. von Neumann: Wahrscheinlichkeitstheoretischer Aufbau der Quantenmechanik
Göttinger Nachrichten **1** (1927) 245-272
in [175] 208-235
167. J. von Neumann: Thermodynamik quantenmechanischer Gesamtheiten
Göttinger Nachrichten **1** (1927) 273-291
in [175] 236-254
168. J. von Neumann *Mathematische Grundlagen der Quantenmechanik*, Springer Verlag, Heidelberg, 1932
169. J. von Neumann: Letter to G. Birkhoff, November 3, 1935
170. J. von Neumann: On an algebraic generalization of the quantum mechanical formalism (Part I)
Mathematical Sbornik **1** (1936) 415-484
in [177] 492-561
171. J. von Neumann: Quantum logics (strict- and probability-logics)
unfinished manuscript, reviewed by A.H. Taub in [178] 195-197
172. J. von Neumann: On rings of operators III
Annals of Mathematics **41** (1940) 94-161
in [177] 6-119
173. J. von Neumann: *Continuous Geometry*, Princeton University Press, Princeton, 1960
174. J. von Neumann: Continuous geometries with transition probability
Memoirs of the American Mathematical Society **34** No. 252 (1981) 1-210.
175. J. von Neumann: *Collected Works Vol. I. Logic, Theory of Sets and Quantum Mechanics*, A.H. Taub (ed.), Pergamon Press, 1962
176. J. von Neumann: *Collected Works Vol. II. Operators, Ergodic Theory and Almost Periodic Functions in a Group*, A.H. Taub (ed.), Pergamon Press, 1962
177. J. von Neumann: *Collected Works Vol. III. Rings of Operators*, A.H. Taub (ed.), Pergamon Press, 1961
178. J. von Neumann: *Collected Works Vol. IV. Continuous Geometry and Other Topics*, A.H. Taub (ed.), Pergamon Press, 1961

INDEX

algebra 78
    AFD 89
    C^*- 78, 79, 146
    commutative 41, 45,
    Banach 78
    Boolean 33, 38, 41, 52-53, 61, 65, 67-68, 103-104, 217
    Jordan 5, 82, 151, 159
    Jordan-Banach 82
    partial 40, 41, 43, 53, 163
        Boolean 29, 42-43
    quasilocal 164, 173
    Tarski-Lindenbaum 4, 61, 67
    Von Neumann 3, 4-5, 8, 59, 77, 81-82, 84, 86, 90-91, 94, 97, 99, 100, 115, 129
    Type I 85
    Type I_n 85
    Type I_∞ 85
    Type II_1 4-5, 85, 88, 200
    Type II_∞ 85
    Type III 85, 176
    type of 85
    UHF 153

cause
    common 3, 6, 7, 189, 215-221, 223-225

commutant 80, 82

complementary 3-4, 27, 198
    operators 16-18, 159

complementarity 17-18, 20, 27, 160

conditional 5, 119-121, 124, 134, 136
    counterfactual 5, 119, 124-126
    expectation 5, 88-89, 119, 129-133, 135, 137, 152-154, 159-160, 199
    Mittelstaedt 5, 122-123, 126, 134-135
    probability 5, 127, 129, 216
    quantum 5, 119, 121, 123-124, 134-135, 137

cone 175
    cyclicity 175
    double 175, 187
        tangent 181, 193

correlation 6, 221
    Bell 6, 171, 180, 184, 188, 190, 224
    probabilistic 215
    statistical 7
    superluminal 7, 188, 215

dimension 83, 114
    function 31-34, 84, 87, 90, 98, 104, 112
    relative 84
    theory 4, 101

dispersion 5, 22, 153
    overall 156

factor 81
    type II_1 4, 85-89, 103-104, 112-113, 115, 117

filter 37, 51

maximal 37-38
prime 37-40, 51, 53
principal 37
proper 37-38, 52
ultra 37

GNS
construction 79
representation 173
triplet 182

hidden variables 3, 139, 143, 161
approximate 155

hidden theory 139, 142, 144, 146, 148, 150-153, 161, 164
entropic 157-158
local 166-167

homomorphism
Boolean algebra 53, 169, 218
Jordan 151-152, 157-158
lattice 52-53
partial algebra 42-43, 45, 53, 163-164

ideal 37, 39, 149
maximal 38
prime 38, 51-53
principal 38
proper 37-38

independence 3, 6-7, 177, 190-191, 212
C^*-177-179, 190-191, 199-200, 202-204, 209-210, 212, 215, 223
counterfactual 7-8, 192, 207
logical 6, 192-193, 195-196, 198, 200, 202, 213
semantic 6, 193, 195-196
statistical 6, 192, 213, 222
W^*-178-179, 191-192, 210, 223

$\Rightarrow$- 194-197

inequality
Bell's 6, 162, 171, 179-180, 182-183, 187-189, 190, 193, 211, 223-225
Cauchy-Schwartz 146, 150

lattice 4, 29, 85, 115
atomic 4, 30, 34, 45
atomistic 45
Boolean 29
bounded 29-30
complete 30, 46, 82
completely atomistic 30-31, 45
distributive 31, 33, 38, 58
Hilbert 4
homomorphism 29, 39
Jauch-Piron 42, 202-203
modular 31, 34, 48, 50, 99, 103, 115
orthocomplemented 33, 35, 115, 121, 122
orthomodular 4, 34, 45, 53, 73, 122
polinom 32
Von Neumann 3, 6, 77, 90, 116

logic 2, 61, 68-70, 73-75, 90, 104, 114, 137
quantum 1, 2, 5-9, 11, 43, 50, 53, 58-59, 74, 77, 103, 116, 199

map
positive linear 135, 146
Einstein local 165

measure 34
complex 14
projection valued 14-15, 21
spectral 15

operator 15, 22, 77-78
    density 21
    Hilbert-Schmidt 20
    position 12, 197
    momentum 13, 15, 197
    statistical 107-111
    trace class 20-21
    finite range 23

probability 5, 21, 104-111, 114, 117, 127-128, 130, 134, 140
    a priori 103, 106, 108-109, 111-113, 131
    conditional 5, 127, 129, 216
    ensemble interpretation of 107
    relative 108

projection 14, 54, 69, 90, 130-133, 135, 137
    finite 84
    infinite 208

screening off 217, 225
    principle 219-220, 223

semantic 4
    independence 6, 193, 195-196
semantics
    possible world 5, 119, 125-126, 135, 205
    Stalnaker 7, 124, 126

spectral measure 15

spectrum 12, 15, 18
    condition 173-174

split property 185

state 11, 21, 42, 79, 130, 136, 142, 149
    clustering 176

dispersion-free 22, 53, 146
faithful 135, 222
Jauch-Piron 42, 51, 202, 204
locally normal 166
monotone positive 147
product 223
pure 145, 150, 158, 180
tracial 132, 199
vacuum 7, 173, 175-177, 215, 221

strict locality 178, 191-192, 223

theorem
    Cantor-Bernstein 83, 94
    comparison 86, 99
    double commutant 80-81
    Gelfand-Naimark-Segal 79
    Gleason's 3, 21, 27, 51-52, 147
    Heisenberg 3, 23
    Kochen-Specker 60
    Mittelstaedt 122
    Paley-Wiener 20
    polar decomposition 91
    Reeh-Schlieder 168, 175, 221-222
    Riesz-Thorun 24, 28
    Schlieder-Roos 177, 213, 221
    spectral 15
    Von Neumann 23

topology
    strong (operator) 79
    ultraweak (operator) 79
    uniform 78
    weak (operator) 79
    w^* 165
trace 21, 77, 86, 97-98, 103, 110, 112-114
    finite 86
    center valued 87, 90, 97-98

uncertainty 5, 144-145, 148

relation 23
entropic 23, 156
  relation 4, 18, 22, 24, 27
statistical 5, 6, 168

wedge 174
  right 174
  left 174

Fundamental Theories of Physics

Series Editor: Alwyn van der Merwe, *University of Denver, USA*

1. M. Sachs: *General Relativity and Matter.* A Spinor Field Theory from Fermis to Light-Years. With a Foreword by C. Kilmister. 1982 ISBN 90-277-1381-2
2. G.H. Duffey: *A Development of Quantum Mechanics.* Based on Symmetry Considerations. 1985 ISBN 90-277-1587-4
3. S. Diner, D. Fargue, G. Lochak and F. Selleri (eds.): *The Wave-Particle Dualism.* A Tribute to Louis de Broglie on his 90th Birthday. 1984 ISBN 90-277-1664-1
4. E. Prugovečki: *Stochastic Quantum Mechanics and Quantum Spacetime.* A Consistent Unification of Relativity and Quantum Theory based on Stochastic Spaces. 1984; 2nd printing 1986 ISBN 90-277-1617-X
5. D. Hestenes and G. Sobczyk: *Clifford Algebra to Geometric Calculus.* A Unified Language for Mathematics and Physics. 1984
 ISBN 90-277-1673-0; Pb (1987) 90-277-2561-6
6. P. Exner: *Open Quantum Systems and Feynman Integrals.* 1985 ISBN 90-277-1678-1
7. L. Mayants: *The Enigma of Probability and Physics.* 1984 ISBN 90-277-1674-9
8. E. Tocaci: *Relativistic Mechanics, Time and Inertia.* Translated from Romanian. Edited and with a Foreword by C.W. Kilmister. 1985 ISBN 90-277-1769-9
9. B. Bertotti, F. de Felice and A. Pascolini (eds.): *General Relativity and Gravitation.* Proceedings of the 10th International Conference (Padova, Italy, 1983). 1984
 ISBN 90-277-1819-9
10. G. Tarozzi and A. van der Merwe (eds.): *Open Questions in Quantum Physics.* 1985
 ISBN 90-277-1853-9
11. J.V. Narlikar and T. Padmanabhan: *Gravity, Gauge Theories and Quantum Cosmology.* 1986 ISBN 90-277-1948-9
12. G.S. Asanov: *Finsler Geometry, Relativity and Gauge Theories.* 1985
 ISBN 90-277-1960-8
13. K. Namsrai: *Nonlocal Quantum Field Theory and Stochastic Quantum Mechanics.* 1986 ISBN 90-277-2001-0
14. C. Ray Smith and W.T. Grandy, Jr. (eds.): *Maximum-Entropy and Bayesian Methods in Inverse Problems.* Proceedings of the 1st and 2nd International Workshop (Laramie, Wyoming, USA). 1985 ISBN 90-277-2074-6
15. D. Hestenes: *New Foundations for Classical Mechanics.* 1986
 ISBN 90-277-2090-8; Pb (1987) 90-277-2526-8
16. S.J. Prokhovnik: *Light in Einstein's Universe.* The Role of Energy in Cosmology and Relativity. 1985 ISBN 90-277-2093-2
17. Y.S. Kim and M.E. Noz: *Theory and Applications of the Poincaré Group.* 1986
 ISBN 90-277-2141-6
18. M. Sachs: *Quantum Mechanics from General Relativity.* An Approximation for a Theory of Inertia. 1986 ISBN 90-277-2247-1
19. W.T. Grandy, Jr.: *Foundations of Statistical Mechanics.*
 Vol. I: *Equilibrium Theory.* 1987 ISBN 90-277-2489-X
20. H.-H von Borzeszkowski and H.-J. Treder: *The Meaning of Quantum Gravity.* 1988
 ISBN 90-277-2518-7
21. C. Ray Smith and G.J. Erickson (eds.): *Maximum-Entropy and Bayesian Spectral Analysis and Estimation Problems.* Proceedings of the 3rd International Workshop (Laramie, Wyoming, USA, 1983). 1987 ISBN 90-277-2579-9

Fundamental Theories of Physics

22. A.O. Barut and A. van der Merwe (eds.): *Selected Scientific Papers of Alfred Landé.* [*1888-1975*]. 1988   ISBN 90-277-2594-2
23. W.T. Grandy, Jr.: *Foundations of Statistical Mechanics.*
    Vol. II: *Nonequilibrium Phenomena.* 1988   ISBN 90-277-2649-3
24. E.I. Bitsakis and C.A. Nicolaides (eds.): *The Concept of Probability.* Proceedings of the Delphi Conference (Delphi, Greece, 1987). 1989   ISBN 90-277-2679-5
25. A. van der Merwe, F. Selleri and G. Tarozzi (eds.): *Microphysical Reality and Quantum Formalism, Vol. 1.* Proceedings of the International Conference (Urbino, Italy, 1985). 1988   ISBN 90-277-2683-3
26. A. van der Merwe, F. Selleri and G. Tarozzi (eds.): *Microphysical Reality and Quantum Formalism, Vol. 2.* Proceedings of the International Conference (Urbino, Italy, 1985). 1988   ISBN 90-277-2684-1
27. I.D. Novikov and V.P. Frolov: *Physics of Black Holes.* 1989   ISBN 90-277-2685-X
28. G. Tarozzi and A. van der Merwe (eds.): *The Nature of Quantum Paradoxes.* Italian Studies in the Foundations and Philosophy of Modern Physics. 1988
    ISBN 90-277-2703-1
29. B.R. Iyer, N. Mukunda and C.V. Vishveshwara (eds.): *Gravitation, Gauge Theories and the Early Universe.* 1989   ISBN 90-277-2710-4
30. H. Mark and L. Wood (eds.): *Energy in Physics, War and Peace.* A Festschrift celebrating Edward Teller's 80th Birthday. 1988   ISBN 90-277-2775-9
31. G.J. Erickson and C.R. Smith (eds.): *Maximum-Entropy and Bayesian Methods in Science and Engineering.*
    Vol. I: *Foundations.* 1988   ISBN 90-277-2793-7
32. G.J. Erickson and C.R. Smith (eds.): *Maximum-Entropy and Bayesian Methods in Science and Engineering.*
    Vol. II: *Applications.* 1988   ISBN 90-277-2794-5
33. M.E. Noz and Y.S. Kim (eds.): *Special Relativity and Quantum Theory.* A Collection of Papers on the Poincaré Group. 1988   ISBN 90-277-2799-6
34. I.Yu. Kobzarev and Yu.I. Manin: *Elementary Particles. Mathematics, Physics and Philosophy.* 1989   ISBN 0-7923-0098-X
35. F. Selleri: *Quantum Paradoxes and Physical Reality.* 1990   ISBN 0-7923-0253-2
36. J. Skilling (ed.): *Maximum-Entropy and Bayesian Methods.* Proceedings of the 8th International Workshop (Cambridge, UK, 1988). 1989   ISBN 0-7923-0224-9
37. M. Kafatos (ed.): *Bell's Theorem, Quantum Theory and Conceptions of the Universe.* 1989   ISBN 0-7923-0496-9
38. Yu.A. Izyumov and V.N. Syromyatnikov: *Phase Transitions and Crystal Symmetry.* 1990   ISBN 0-7923-0542-6
39. P.F. Fougère (ed.): *Maximum-Entropy and Bayesian Methods.* Proceedings of the 9th International Workshop (Dartmouth, Massachusetts, USA, 1989). 1990
    ISBN 0-7923-0928-6
40. L. de Broglie: *Heisenberg's Uncertainties and the Probabilistic Interpretation of Wave Mechanics.* With Critical Notes of the Author. 1990   ISBN 0-7923-0929-4
41. W.T. Grandy, Jr.: *Relativistic Quantum Mechanics of Leptons and Fields.* 1991
    ISBN 0-7923-1049-7
42. Yu.L. Klimontovich: *Turbulent Motion and the Structure of Chaos.* A New Approach to the Statistical Theory of Open Systems. 1991   ISBN 0-7923-1114-0

Fundamental Theories of Physics

43. W.T. Grandy, Jr. and L.H. Schick (eds.): *Maximum-Entropy and Bayesian Methods.* Proceedings of the 10th International Workshop (Laramie, Wyoming, USA, 1990). 1991 ISBN 0-7923-1140-X
44. P.Pták and S. Pulmannová: *Orthomodular Structures as Quantum Logics.* Intrinsic Properties, State Space and Probabilistic Topics. 1991 ISBN 0-7923-1207-4
45. D. Hestenes and A. Weingartshofer (eds.): *The Electron.* New Theory and Experiment. 1991 ISBN 0-7923-1356-9
46. P.P.J.M. Schram: *Kinetic Theory of Gases and Plasmas.* 1991 ISBN 0-7923-1392-5
47. A. Micali, R. Boudet and J. Helmstetter (eds.): *Clifford Algebras and their Applications in Mathematical Physics.* 1992 ISBN 0-7923-1623-1
48. E. Prugovečki: *Quantum Geometry.* A Framework for Quantum General Relativity. 1992 ISBN 0-7923-1640-1
49. M.H. Mac Gregor: *The Enigmatic Electron.* 1992 ISBN 0-7923-1982-6
50. C.R. Smith, G.J. Erickson and P.O. Neudorfer (eds.): *Maximum Entropy and Bayesian Methods.* Proceedings of the 11th International Workshop (Seattle, 1991). 1993
ISBN 0-7923-2031-X
51. D.J. Hoekzema: *The Quantum Labyrinth.* 1993 ISBN 0-7923-2066-2
52. Z. Oziewicz, B. Jancewicz and A. Borowiec (eds.): *Spinors, Twistors, Clifford Algebras and Quantum Deformations.* Proceedings of the Second Max Born Symposium (Wrocław, Poland, 1992). 1993 ISBN 0-7923-2251-7
53. A. Mohammad-Djafari and G. Demoment (eds.): *Maximum Entropy and Bayesian Methods.* Proceedings of the 12th International Workshop (Paris, France, 1992). 1993
ISBN 0-7923-2280-0
54. M. Riesz: *Clifford Numbers and Spinors* with Riesz' Private Lectures to E. Folke Bolinder and a Historical Review by Pertti Lounesto. E.F. Bolinder and P. Lounesto (eds.). 1993 ISBN 0-7923-2299-1
55. F. Brackx, R. Delanghe and H. Serras (eds.): *Clifford Algebras and their Applications in Mathematical Physics.* Proceedings of the Third Conference (Deinze, 1993) 1993
ISBN 0-7923-2347-5
56. J.R. Fanchi: *Parametrized Relativistic Quantum Theory.* 1993 ISBN 0-7923-2376-9
57. A. Peres: *Quantum Theory: Concepts and Methods.* 1993 ISBN 0-7923-2549-4
58. P.L. Antonelli, R.S. Ingarden and M. Matsumoto: *The Theory of Sprays and Finsler Spaces with Applications in Physics and Biology.* 1993 ISBN 0-7923-2577-X
59. R. Miron and M. Anastasiei: *The Geometry of Lagrange Spaces: Theory and Applications.* 1994 ISBN 0-7923-2591-5
60. G. Adomian: *Solving Frontier Problems of Physics: The Decomposition Method.* 1994
ISBN 0-7923-2644-X
61. B.S. Kerner and V.V. Osipov: *Autosolitons.* A New Approach to Problems of Self-Organization and Turbulence. 1994 ISBN 0-7923-2816-7
62. G.R. Heidbreder (ed.): *Maximum Entropy and Bayesian Methods.* Proceedings of the 13th International Workshop (Santa Barbara, USA, 1993) 1996 ISBN 0-7923-2851-5
63. J. Peřina, Z. Hradil and B. Jurčo: *Quantum Optics and Fundamentals of Physics.* 1994
ISBN 0-7923-3000-5
64. M. Evans and J.-P. Vigier: *The Enigmatic Photon.* Volume 1: The Field $B^{(3)}$. 1994
ISBN 0-7923-3049-8
65. C.K. Raju: *Time: Towards a Constistent Theory.* 1994 ISBN 0-7923-3103-6
66. A.K.T. Assis: *Weber's Electrodynamics.* 1994 ISBN 0-7923-3137-0

Fundamental Theories of Physics

67. Yu. L. Klimontovich: *Statistical Theory of Open Systems.* Volume 1: A Unified Approach to Kinetic Description of Processes in Active Systems. 1995
ISBN 0-7923-3199-0; Pb: ISBN 0-7923-3242-3
68. M. Evans and J.-P. Vigier: *The Enigmatic Photon.* Volume 2: Non-Abelian Electrodynamics. 1995 ISBN 0-7923-3288-1
69. G. Esposito: *Complex General Relativity.* 1995 ISBN 0-7923-3340-3
70. J. Skilling and S. Sibisi (eds.): *Maximum Entropy and Bayesian Methods.* Proceedings of the Fourteenth International Workshop on Maximum Entropy and Bayesian Methods. 1996 ISBN 0-7923-3452-3
71. C. Garola and A. Rossi (eds.): *The Foundations of Quantum Mechanics – Historical Analysis and Open Questions.* 1995 ISBN 0-7923-3480-9
72. A. Peres: *Quantum Theory: Concepts and Methods.* 1995 (see for hardback edition, Vol. 57) ISBN Pb 0-7923-3632-1
73. M. Ferrero and A. van der Merwe (eds.): *Fundamental Problems in Quantum Physics.* 1995 ISBN 0-7923-3670-4
74. F.E. Schroeck, Jr.: *Quantum Mechanics on Phase Space.* 1996 ISBN 0-7923-3794-8
75. L. de la Peña and A.M. Cetto: *The Quantum Dice.* An Introduction to Stochastic Electrodynamics. 1996 ISBN 0-7923-3818-9
76. P.L. Antonelli and R. Miron (eds.): *Lagrange and Finsler Geometry.* Applications to Physics and Biology. 1996 ISBN 0-7923-3873-1
77. M.W. Evans, J.-P. Vigier, S. Roy and S. Jeffers: *The Enigmatic Photon.* Volume 3: Theory and Practice of the $B^{(3)}$ Field. 1996 ISBN 0-7923-4044-2
78. W.G.V. Rosser: *Interpretation of Classical Electromagnetism.* 1996
ISBN 0-7923-4187-2
79. K.M. Hanson and R.N. Silver (eds.): *Maximum Entropy and Bayesian Methods.* 1996
ISBN 0-7923-4311-5
80. S. Jeffers, S. Roy, J.-P. Vigier and G. Hunter (eds.): *The Present Status of the Quantum Theory of Light.* Proceedings of a Symposium in Honour of Jean-Pierre Vigier. 1997
ISBN 0-7923-4337-9
81. M. Ferrero and A. van der Merwe (eds.): *New Developments on Fundamental Problems in Quantum Physics.* 1997 ISBN 0-7923-4374-3
82. R. Miron: *The Geometry of Higher-Order Lagrange Spaces.* Applications to Mechanics and Physics. 1997 ISBN 0-7923-4393-X
83. T. Hakioğlu and A.S. Shumovsky (eds.): *Quantum Optics and the Spectroscopy of Solids.* Concepts and Advances. 1997 ISBN 0-7923-4414-6
84. A. Sitenko and V. Tartakovskii: *Theory of Nucleus.* Nuclear Structure and Nuclear Interaction. 1997 ISBN 0-7923-4423-5
85. G. Esposito, A.Yu. Kamenshchik and G. Pollifrone: *Euclidean Quantum Gravity on Manifolds with Boundary.* 1997 ISBN 0-7923-4472-3
86. R.S. Ingarden, A. Kossakowski and M. Ohya: *Information Dynamics and Open Systems.* Classical and Quantum Approach. 1997 ISBN 0-7923-4473-1
87. K. Nakamura: *Quantum versus Chaos.* Questions Emerging from Mesoscopic Cosmos. 1997 ISBN 0-7923-4557-6
88. B.R. Iyer and C.V. Vishveshwara (eds.): *Geometry, Fields and Cosmology.* Techniques and Applications. 1997 ISBN 0-7923-4725-0

Fundamental Theories of Physics

89. G.A. Martynov: *Classical Statistical Mechanics.* 1997   ISBN 0-7923-4774-9
90. M.W. Evans, J.-P. Vigier, S. Roy and G. Hunter (eds.): *The Enigmatic Photon.* Volume 4: New Directions. 1998   ISBN 0-7923-4826-5
91. M. Rédei: *Quantum Logic in Algebraic Approach.* 1998   ISBN 0-7923-4903-2
92. S. Roy: *Statistical Geometry and Applications to Microphysics and Cosmology.* 1998   ISBN 0-7923-4907-5